TAKEN ON CRETE 1941

Cpl James Roberts' War

Gary Hocking

AUTHOR'S NOTE

James Roberts was my mother Lorna's brother. He wouldn't talk much about his war but over many years I pecked away at him and he told me much about it. The book is a true story from what he told me, what my grandparents told me, what my mother told me and the Australian Military Archives.

We had a strong bond and I tell this story so that history and our families can remember and learn about his terrible but triumphant ordeal.

I truly hope I have done well in honouring him and his service to Australia.

Gary Hocking
2025

CHAPTER 1 — NEWCASTLE, NSW.

The house in Alfred Street stood one block back from the beach, close enough that the sound of the surf travelled up the street at night. In summer, salt dried on the window frames and left them rough to the touch. The house was weatherboard, narrow, with rooms arranged one behind the other and a steep stair that creaked no matter how carefully it was climbed. There was no part of the house where a person could be alone for long. Voices and footsteps moved through the boards and made their way into every room.

Dick was the eldest of the children. He resembled his father around the eyes, though he had his mother's height and stood very still when thinking, and at twenty-two he had learned how to work without needing to be told. Owen Roberts worked nights at The Newcastle Herald as a linotype operator. He would return home at sunrise with ink embedded deep in the grooves of his fingers. The newspaper smell never left his clothes despite washing. During the day he slept behind the closed door of the front room, the rest of the household moving around him with a natural quietness that came from years of living together in close space.

Lilian ran the home with steady hands, moving from stove to table to laundry and back again, her movements efficient after years of managing a crowded household. With five daughters and one son, the house was never quiet, never still. Washing hung on the line every day. Meals required careful planning and precise timing. She rarely raised her voice. She did not need to. One look or a single word set things right again.

Jean and Lorna were closest to Dick in age, already young women at nineteen and seventeen. They were quick with laughter and could argue one moment and sit together peacefully the next. Dorrie was fifteen, practical and steady, often helping their mother manage the younger ones. Shirley, at eleven, had a quick smile and was never far from Meryl, the youngest at nine, who followed their mother from room to room and lifted pots and spoons with all the seriousness of work.

On Saturday mornings, Dick would sometimes find Jean, Lorna, and Dorrie at the kitchen table, heads bent over the newspaper, reading aloud the names of boys from Newcastle who had enlisted. Jean would pause at certain names. "That's Betty Marsh's brother," she'd say. Or, "Didn't he work at the steelworks?" Lorna usually had her own stories to add, who had signed up with whom, which mothers were trying to talk their sons out of it. Dorrie mostly listened, her finger tracing the columns of names as if memorizing them.

The lists had been growing since the previous September, when the war began in earnest.

Dick worked as a clerk at the shipping office near the wharves. The work was orderly: manifests, bills of lading, cargo lists written in careful script. He recorded the weight of wool bales and the destination of coal shipments. The office smelled of paper and tobacco and the harbour beyond the windows. Through the glass he could see the cranes moving, the ships tied up at the docks, their hulls dark against the water. Men shouted to each other across the wharves in voices that carried over the sound of engines and winches.

He was good at his work. His figures were neat, his filing methodical. He rarely made mistakes that needed correcting. On Friday afternoons, when Mr. Peterson distributed the pay envelopes, Dick's always contained a small bonus for accuracy. Steady employment that could lead to advancement if a man wanted it.

The office had changed in recent months. Two of the younger clerks had already enlisted, their desks cleared and reassigned. Mr. Peterson, who managed the schedules, had served in the last war and sometimes stood at the window watching the harbour with an expression that suggested he was seeing something other than what was there. The cargo had changed too. More government requisitions now, less private trade. Ships that used to carry wool and wheat were being fitted for other purposes, though no one spoke directly about what those purposes were.

One afternoon in early September, Dick sat at his desk copying figures from a stack of delivery dockets. The work required attention but not much thought, and his hand moved steadily across the page. Through the window he could see a ship being loaded, the crane swinging its cargo in slow arcs. The foreman was shouting instructions that barely reached the office through the glass.

Mr. Peterson walked past and paused at Dick's desk. He was a thin man with grey hair who stood as if his back bothered him.

"Your work's always neat, Roberts," he said. "Never have to check it twice."

"Thank you, sir."

Peterson looked out the window at the harbour. A long silence stretched between them. Then: "My boy was about your age when he went. Seventeen. Lied about it to get in." Another pause. "He didn't come back from France."

Dick set down his pen. Peterson had never mentioned his son before.

"I'm not telling you what to do," Peterson said, still looking at the harbour. "Just saying there's no shame in staying if that's what you choose. Someone's got to keep things running here."

He walked away before Dick could respond. Dick looked down at the half-finished page of figures and found he had lost his place. He started the line again, his handwriting as steady as before, but his thoughts had moved elsewhere.

By mid-1940, the talk of war had become impossible to avoid. Hitler had taken Poland, then Norway, then France. The maps in the newspapers showed Europe darkening country by country. England stood alone now. The words "if England falls" were spoken more often, though never quite finished. Everyone knew what came after those words.

At the shipping office, older men who had served in the last war spoke about France and Gallipoli, their voices taking on a particular tone when they mentioned certain places. They did not glamorize it. They simply spoke of it as something they had done and survived. In the hotels along Hunter Street, younger men, men Dick's age, spoke with a mixture of certainty and restlessness. Some said Australia was too far away to matter, that the war would be over before it reached the Pacific. Others said that was fool's thinking, that Japan was watching everything that happened in Europe, waiting for its moment.

Dick listened to these conversations without adding much to them. He had never been good with speeches or grand statements. He understood what was at stake, even if he had no words that would satisfy a recruiting poster.

On warm afternoons, when the work at the shipping office was done, Dick often walked to the beach. The sand was pale and fine and the slope of the shore caught the sun in broad, bright arcs. He would stand with his hands in his pockets and look out across the water. The ocean had always been there, as much a part of Newcastle as the coal dust and the steel mills.

He had learned to swim here when he was six or seven, his father walking beside him into the shallows on a summer morning before the crowds arrived. Owen had been patient, holding Dick steady in the water until the boy found his balance and kicked out on his own. "Don't fight it," his father had said. "Just move with it." Dick remembered the feeling of the water holding him up, the surprise of discovering he could float.

Now he stood on the same beach and thought about distance. England was on the other side of the world, twelve thousand miles away. It seemed impossible that what happened there should matter here, on this beach, in this town where the trams ran on time and the ships still came and went. But the newspapers said otherwise. The wireless said otherwise. And the men who had fought in the last war said otherwise.

The shipping office, with its neat rows of figures and predictable routines, had begun to feel smaller. It was a good job. Steady. A man could build a life around it. But lately the walls pressed closer while the world outside grew larger and more urgent.

He turned and walked back up the beach toward the street. The sun was lowering, and the light caught the windows of the houses along the esplanade, turning them gold.

That evening, Dick met Eileen at the corner near the post office. She was waiting under the awning of the pharmacy, her handbag held in both hands. She smiled when he saw him, and they walked together toward the park near the railway line where couples often sat in the evenings.

Eileen worked at the telephone exchange, her voice calm and precise as she connected calls from one end of town to the other. She had dark hair that she wore pinned back, and when she listened she tilted her head in a way that made people feel heard. They had been stepping out together for nearly a year, long enough that both families expected something would be decided before too much more time passed.

They sat on a bench under a fig tree. Dick heard the trains shunting in the yards, the clanking carrying across open ground. For a while they talked about small things. Her day at the exchange, his work at the shipping office, a film showing at the Lyric that everyone said was worth seeing.

Then Eileen said, "Tom Brennan's mother told mine that he's enlisted."

Dick nodded. "He and some of the others. Last week."

"Are you thinking about it?"

He had known the question would come. Eileen never avoided difficult conversations. "I am," he said.

She looked down at her hands, turning her handbag. "I thought you might be."

They sat in silence. A tram rattled past on the street beyond the park, its bell clanging twice at the intersection. When it had gone, Eileen's voice was quieter than before.

"If you're going to go, I want us to be engaged first. Before you leave."

His chest tightened. He looked at her hands gripping the handbag, at the careful way she held herself. Marriage had always seemed like something that would happen eventually, when the time was right, when he had saved enough, when everything felt settled.

"I can't do that," he said.

Her head came up. "Why not?"

"Because..." He stopped, tried again. "I don't know what's going to happen over there. When I'll be back." He made himself say it. "If I'll be back."

"That should be for me to decide."

"Maybe. But it's for me to decide too."

She turned away from him, her shoulders stiff. A group of children were playing near the swings, their voices high and bright in the cooling air. When she spoke again, each word was careful, deliberate.

"You think you're being fair by waiting. But you're not. You're just making it so neither of us has anything certain."

The words hit harder than he expected. He looked at his own hands. "If I come back, we'll get engaged then. We'll do it properly. But I can't..." He wouldn't ask her to tie herself to a man who might never walk back through that door. Wouldn't make her into one of those girls who waited years for news that never came.

"You're not asking. I'm offering."

"And I'm saying no."

The words came out harder than he intended. She flinched, a quick tightening around her eyes though her expression did not otherwise change. She stood, one hand smoothing her skirt, the other still gripping her handbag.

"I should get home," she said. "My mother will wonder where I am."

He stood as well, his stomach heavy. "Eileen..."

"It's all right." Her voice was steady but her eyes were

bright. "I understand."

But her tone said otherwise.

They walked back through the park without speaking. At the corner where their paths would diverge, she stopped and looked at him. Her eyes were too bright now, though no tears fell.

"When you go," she said, "write to me. Even if we're not engaged. Write to me anyway."

"I will."

"Promise."

"I promise."

She nodded once, then turned and walked away. He watched until she rounded the corner and disappeared from view. A breach had opened between them that he didn't know how to mend.

He walked home slowly, his hands in his pockets.

One evening a few days later, the sound of boots on the timber verandah brought him outside. Four of his mates were there: Tom Brennan, who worked at the BHP steelworks; Jack Morrison, a carpenter; Bill Hughes from the railway workshops; and Frank Delaney, who had been two years behind Dick at school and now worked at the gasworks. Men he had grown up with, worked beside, played football with on winter Saturdays at Sportsground Number Two.

Tom's hands were stained dark from the steelworks, staining that never quite washed out. Jack still had sawdust on his trousers. Bill stood with his cap in his hands, turning it. Frank, the youngest of them, looked both eager and uncertain.

Their stance had changed. Their clothes were the same as always, but they stood steadier now, more resolved.

"We've signed up," Tom said.

He understood immediately. He had known it was coming, had felt it in the way conversations at work had shifted, in the way his mates had grown quieter when the talk turned to the war.

"All of you?" Dick asked.

"All of us," Jack said. "Last week."

"Bill's missus wasn't happy about it," Tom added. "But she didn't try to stop him."

Bill shrugged. "She knew it was coming. We all did."

He looked at them, these men whose lives had run parallel to his own for as long as he could remember. They had learned to swim at the same beaches, had sat in the same classrooms, had drunk at the same hotels. Them going somewhere without him, it was wrong, though he couldn't say why.

Dick's hand found the verandah rail. "You're not going without me," he said.

The words hung in the air for a moment. Then Tom grinned, the tension breaking. "Thought you might say that."

Frank let out a breath. "Good. It wouldn't be right without you."

They talked for a while longer on the verandah, their voices low so as not to wake Owen. When they might leave, where they might be sent, whether the training would be hard. None of them knew much. None of them had been further than Sydney. Egypt or England or wherever they would be sent existed only as words and maps and the stories of older men.

"What about Eileen?" Tom asked. "She know yet?"

"She knows I've been thinking about it."

"She'll wait," Tom said with certainty. "Girls like that always do."

He did not correct him. There was no point explaining what had passed between them in the park.

After his mates left, Dick went back inside. The house was quiet. His mother was in the kitchen, washing the dinner dishes. She glanced at him as he passed but said nothing. She knew things without being told.

That night, lying awake in the narrow room he had to himself now that he was the only boy in the house, he could hear the ocean through the window. The regular crash and hiss of waves on sand. The shipping office. His sisters. His mother. His father coming home at dawn with ink on his hands. Eileen under the awning of the pharmacy, the way she had looked at him before walking away.

The beach and the town and the life he had always assumed would continue in an unbroken line from what it was now to what it would be when he was his father's age. But the war had arrived even here, even in this quiet street near the sea. And now that his mates had made their decision, there was only one decision he could make.

He tried to picture himself on a ship twelve thousand miles from home, tried to see what lay on the other side of all that water, but the images would not form. There was only the darkness of his room and the sound of the ocean and the knowledge that soon he would be among those men, moving toward a future he could not yet see.

They went to the recruiting office in Hunter Street the next morning. Tom and Jack walked with Dick. Bill had gone the day before, and Frank would come later. The street was busy with delivery carts, trams ringing their bells as they took the corner near the post office, and workers moving toward the docks. The air smelled of coal smoke and bread from the bakery on the corner.

They passed a group of women standing outside the butcher's shop, talking in low voices. One of them looked at Dick and the others and seemed to understand where they were going. She gave a small nod, neither approving nor disapproving, simply acknowledging what was happening all over town.

The entrance to the Government Buildings smelled of old paper and floor polish. Inside, the corridor was dim and cool after the brightness of the street. They waited in line behind men from Hamilton, Lambton, Merewether, Cooks Hill. Some talked quietly. Some said nothing. Dick recognized a few faces from the football grounds, from the hotels, from the wharves.

The mood was not celebratory. There were no bands or flags, no crowds cheering them on. The men in line looked like what they were: ordinary men making a decision that would take them far from home. Some looked eager. Some looked resigned. Most were thinking about all the things they would leave behind.

A young bloke in front of them, who couldn't have

been more than eighteen, kept smoothing his hair with one hand. His knee bounced nervously. His mate beside him, older, maybe a brother, put a hand on his shoulder and the bouncing stopped.

When his turn came, Dick gave his name. James Owen Roberts, though everyone called him Dick. His age, his place of residence, and his occupation. The clerk at the desk filled out the form with mechanical precision, barely looking up. There was no talk of patriotism or sacrifice, no stirring words about duty or empire. The process was routine. The war machine had already begun to move and this was one of its simplest gears.

He answered the questions asked of him. Height, weight, any previous military service, next of kin. He signed where he was told to sign. The clerk stamped the form twice and handed him a paper with a date and a place to report for medical examination.

"Next," the clerk said.

With his enlistment paper folded once in his pocket, Dick walked home along Hunter Street. Tom and Jack had gone their separate ways, each heading to their own homes to tell their own families. He passed the tram depot, the fruit vendors calling out their prices, the wool bales stacked near the railway line, and the smell of coal dust that always hung over the harbour. The world looked the same as it had the day before. The same shops, the same faces, the same haze of smoke from the steel mills. Though he knew it was not the same at all.

When he reached the house, his mother stood in the kitchen doorway with a tea towel over her shoulder. She looked at him and then at the paper in his hand. Her face did not change quickly. She took time to see him as he was now.

"When?" she asked.

"Medical exam next week. Then training, I suppose."

She nodded once. She turned back to the stove where potatoes were boiling for dinner. Her hands moved with the same steady rhythm they always had.

"Your sisters will want to know," she said without turning around. "And we'll need to have a proper dinner tonight. All of us."

"Dad's working tonight."

"Then tomorrow," she said. "When he's awake."

That evening, the family gathered in the kitchen. Lilian had made a roast, something she only did on Sundays or when there was something worth marking. The smell of meat and gravy filled the small house. Jean and Lorna set the table while Dorrie helped with the vegetables. Shirley kept Meryl occupied, away from the hot stove.

When they sat down, Owen was at the head of the table, still in his work clothes, the ink not yet fully scrubbed from his hands. He had slept poorly that day, though no one mentioned it. Dick sat to his right, his sisters arranged along the sides of the table.

For a while they ate in silence, the only sounds were the scrape of forks on plates and the occasional request to pass the salt. Then Jean spoke.

"Tom's mother said he enlisted. And Bill Hughes."

"Jack Morrison too," Dick said.

Jean looked at him. "And you."

"Yes."

Lorna set down her fork carefully. She was seventeen and had always been the quieter of the older girls, the one who thought carefully before speaking. "When will you go?"

"Not sure yet. Medical next week. Training after that. Few months, probably."

"That's not very long," she said quietly.

"No," Dick agreed. "It's not."

Meryl, who was nine and did not fully understand what was happening, asked, "Will you be back for Christmas?"

The adults exchanged glances. Before Dick could answer, Jean reached over and squeezed Meryl's hand. "We'll see, love. We'll see."

A small silence fell. Then Shirley, who had been looking at her plate, spoke up. "Betty Carson's brother went last month and she cried at school for three days straight." She looked at Dick, her eyes serious. "I won't cry like that."

"Shirley," their mother said gently.

"Well I won't. It's not helpful." But her voice wavered.

Dorrie, practical as always, said quietly, "We'll manage here. Don't worry about us."

Something tightened in his throat. "I know you will."

"But you'll write to us?" Shirley asked, recovering herself. "All of us? Not just Mum and Dad?"

"All of you," Dick said. "Though you'll have to take turns writing back or I'll spend all my time reading letters."

This brought small smiles around the table, lightening the moment.

Owen cleared his throat. He had been eating slowly, methodically, his attention on his plate. Now he looked up at Dick. "You told Eileen?"

"I did."

"Is she taking it all right?"

The park. The careful way she had held herself. The brightness in her eyes. "She's managing."

Owen nodded and went back to his meal. He was not a man who asked questions he did not need answered.

After dinner, Jean and Lorna helped clear the plates while their mother made tea. Dorrie supervised Shirley and Meryl with the washing up. Dick stepped outside onto the verandah. The air had cooled, and the crash and hiss of waves was clearer now in the evening quiet. He lit a cigarette and stood looking down the street toward the beach, though he could not see it from here.

His father came out and stood beside him. Owen did not smoke anymore. The doctor had told him to stop years ago. But he seemed comfortable in the presence of smoke.

"Your mother's worried," Owen said after a while.

"I know."

"She won't say it. But she is." Owen was quiet for a moment. "It's hard enough managing five girls. She worries about managing without you here to help."

He had not thought of it that way. That his leaving would create a practical burden as well as an emotional one. But of course it would. He was the only man in the house besides his father, who was gone all night at work. He helped with things that needed strength or height. He kept an eye on things when Owen was sleeping.

"Jean and Lorna are old enough," Dick said. "They'll manage."

"They will," Owen agreed. "But it won't be the same."

Owen was quiet for a long moment. Then he said, "I wanted to go in the last one." He looked out at the darkening street rather than at Dick. "I tried to enlist in seventeen. They knocked me back. Reserved occupation. Needed at the paper." The words came slowly, reluctantly. "Sat here the whole war, setting type about other men fighting."

Dick had not known this. His father rarely spoke about the last war.

"You did what you were told to do," Dick said.

"Maybe." Owen's jaw worked. "But it never sat right."

Another silence. Then Owen placed his hand on Dick's shoulder, his palm warm and heavy. "I'm not saying I want you to go. No father wants that. But I understand why you are."

The hand stayed there a moment longer. "You'll do right," Owen said finally. "You've always known how to do right."

Then he went back inside, leaving Dick alone on the verandah. Dick finished his cigarette and ground it out under his heel. Down the street, Mrs. Peterson brought in her washing. Two boys kicked a football in the vacant lot on the corner. Everything was ordinary and familiar, and Dick found himself noticing it all with a sharpness he had not felt before. The exact angle of the light, the smell of someone's dinner cooking, the sound of a dog barking three houses down.

He went back inside. Jean and Lorna were in the front room, reading. Dorrie was mending something by the lamp. Shirley and Meryl were already upstairs, getting ready for bed. His mother was washing the dishes, her hands moving through the soapy water. His father had gone back to the front room to rest before his shift.

In the doorway, Dick stood and watched them for a moment, these people who made up his world, this house that had held him his entire life. Five sisters, and he was leaving them. In a few months he would be gone, and they would still be here, carrying on as they always had. The older ones would take on more responsibility. The younger ones would grow up without him watching.

Jean looked up and caught his eye. She smiled, a complicated expression that held both pride and sadness. Dorrie, seeing the exchange, came over and squeezed his arm briefly before returning to her mending. No one spoke.

The days that followed were strange. Work at the shipping office continued. Meals with his family. Walks through town. But everything had shifted. On his last day at work, Mr. Peterson shook his hand at the door and said nothing, though his grip was firm and his eyes were wet.

He saw Eileen twice more before he left for training. The first time at the post office, where they ran into each other by accident. Their conversation was polite and careful, full of small talk about the weather and mutual acquaintances. The second time at her request. They walked through the same park where they had sat under the fig tree, but neither of them suggested sitting down. The walk was shorter. When they parted at the corner, she touched his arm briefly and said, "Be safe." That was all. She did not ask again about getting engaged. He did not bring it up. They had made their promises. That he would write. That she would wait, though neither of them had said that part aloud.

The medical examination came and went. He was declared fit. The orders arrived by post: report to Ingleburn on the fifteenth of October. He would have three days to settle his affairs.

On his last Sunday, the family went to church together, something they had not done as a complete group in years. They sat in the pew near the back, and when the minister asked for prayers for the men who were going to war, Dick felt his mother's hand reach over and grip his. She held it through the entire prayer, her fingers strong and steady, and when it was done she let go without saying anything.

That afternoon, Dick walked to the beach one final time. The sand was warm under his feet, and the ocean stretched out before him. He stood at the water's edge and let the waves wash over his boots, not caring that they would need to be cleaned.

His father taught him to swim in these waters. All the summers spent here with his mates. Standing in this exact spot only weeks ago wondering what was going to happen. Now he knew. Or at least he knew the next step.

He turned and walked back up the beach toward the street. He did not look back at the ocean. There would be ocean enough where he was going.

The war had begun for Dick in a quiet street near the sea, in a house that did not ask for speeches, and among people who believed that when something needed doing, it was done. It had begun with understanding that some things, once started, could not be stopped. The world had shifted, and he was moving with it, toward something unknown, toward a future that existed only in the distance between here and there.

In three days he would board a train for Ingleburn. In a few months he would board a ship for somewhere else. And Newcastle, with its beaches and its coal dust and its weatherboard houses, would become the place he had been, rather than the place he was.

But not yet. For now, he walked home through the familiar streets, past the houses he had known all his life, toward the people who were waiting for him. For now, he was still here, still home, still the person he had always been.

For now.

CHAPTER 2 — INGLEBURN

The train rattled south through the afternoon, carrying two hundred men away from everything they knew. Dick sat by the window, Tom beside him, watching Newcastle disappear behind them. The harbour, the steelworks, the smoke that always hung over the town. Then suburban sprawl gave way to bush, small stations where people on platforms stared at the soldiers passing through.

No one said much. Some men slept. Others played cards on their packs. Frank kept looking back the way they'd come, as if he could still see home if he tried hard enough.

By evening they reached Ingleburn.

The camp sat on open ground southwest of Sydney where the earth turned to red dust under the sun. Rows of huts were lined in straight formation, each with bunks set close together and a small stove at one end. The air smelled of canvas, boot polish, and the sharp tang of oil from the workshop where rifles were cleaned. Beyond the perimeter fence, gum trees stood in scattered groups, their bark peeling in long strips that curled in the heat.

The sergeants were waiting. They divided the men into sections, assigned hut numbers, and sent them off to collect uniforms and equipment. No one asked why they had come. The army already knew what it needed from them.

Hut 7 held sixteen men. Tom, Jack, Bill, and Frank were all there, assigned to the same section. The bunks were narrow iron frames with thin mattresses that smelled of disinfectant. Each man was given a footlocker for his belongings and a space on the wall for his rifle when it was issued.

The man in the bunk beside Dick was called Mick Dawson. He was from Goulburn, broad through the chest and shoulders, with hands scarred from heavy work. When he shook Dick's hand, his grip was firm but brief.

"You play football?" Mick asked.

"Used to. Newcastle comp."

"Thought so. You've got the build for it."

On Dick's other side was Bluey Patterson. He was thin and wiry, with red hair that explained the nickname. Within five minutes he had introduced himself to everyone in the hut and started offering opinions on everything from the quality of the mattresses to the likely competence of the sergeants.

"These bunks are harder than church pews," he said. "Still, better than sleeping rough. Any of you blokes seen rats yet?"

No one had.

"Good. That's something, anyway."

Tom claimed the bunk across from Dick. Jack was already organizing his footlocker with carpenter's precision. Bill stretched out with his cap over his face. Frank sat watching everything with the uncertain expression of someone still working out what he'd gotten himself into.

"Right then," Tom said, looking around at them. "Here we are."

The kit bag came open easily. Dick folded his civilian clothes at the bottom. The letter from his mother tucked into the corner. A photograph of his family standing in front of the house on Alfred Street. He looked at it for a moment before placing it face-down in the locker.

Reveille sounded before dawn on the first morning. The bugle call cut through sleep like a blade. Dick swung out of his bunk. Around him, men fumbled for clothes in the dim light.

Bluey was already up, his uniform half-buttoned. "That's it then, is it? They just blow that bloody horn and we're meant to jump?" But he was moving as he talked.

Outside, the sky was grey with the approaching dawn. The air was cold and sharp, and their breath came out in white clouds. Forming up in rough lines on the parade ground, they waited while the sergeants walked among them, straightening the formation with brief commands.

Sergeant Harris was in charge of their section. He was perhaps thirty-five, with a face carved from stone. He had served in the last war. It showed in the way he stood, in the economy of his movements.

He walked the length of their line without speaking, his eyes moving from man to man. When he reached the end, he turned and faced them.

"Some of you think you know what's coming," he said. "You don't. But you'll learn. That's what you're here for." He paused. "Right now you're soft. You're slow. You're individuals. My job is to fix that. Your job is to let me."

Breakfast was porridge with the consistency of paste, bread going stale, tea that tasted of tin. Dick ate standing at a long table in the mess hall, shoulder to shoulder with men he didn't know.

"Not exactly your mum's cooking," Tom said.

"Not exactly."

After breakfast they assembled for drill. Marching, turning, breaking ranks and forming again. Left face, right face, about face. The movements were awkward at first, men bumping into each other or turning the wrong way. Sergeant Harris walked among them, his voice patient but firm. "Again."

The morning sun climbed higher. Red dust rose under their boots, coating their uniforms, getting into their mouths and noses. Sweat ran down Dick's back. His legs ached from the constant standing at attention, the quick pivots on the heel.

Beside him, Frank stumbled during an about-face, nearly colliding with the man behind him. Sergeant Harris was there immediately.

"Delaney. Where are you going?"

"Sorry, Sergeant. Lost my footing."

"Your footing is right where you left it. Find it and stay there."

Frank's face went red but he nodded and got back into position.

Through repetition, the movements became automatic. Their bodies learned what their minds still struggled to remember. When Harris called the commands, they responded as one, feet hitting the ground in unison.

The .303 rifles were issued in the second week. When the armourer handed Dick the rifle, the weight surprised him. The Lee-Enfield smelled of gun oil and metal. Each rifle had a serial number, connecting Dick's name to this particular weapon. From now on, this rifle was his responsibility.

Holding it, carrying it, presenting it. The rifle had parts, and each part had a name. Bolt, magazine, barrel, stock, sling, foresight, backsight. Stripping it down and reassembling it, first slowly with the manual, then faster, then in the dark with only their hands.

At night in the hut, the rifle came apart and went back together by feel, Dick's hands moving through the sequence until it became muscle memory.

"You're going to marry that thing, Roberts," Bluey said one evening. "I've seen men less devoted to their wives."

"Just keeping it clean."

Mick, cleaning his own rifle across the aisle, looked up. "Let him be. When we're out there and your rifle jams because you've got sand in the bolt, Roberts will be the one still shooting."

"Point taken," Bluey said.

On the range, they learned to shoot. The first time Dick squeezed the trigger, the recoil surprised him. A solid punch against his shoulder. The crack of the shot echoed across the range. Cordite filled his nose, acrid and sharp. Down the range, the target remained unmarked.

"Flinched," Sergeant Harris said, walking behind the line. "You're anticipating the recoil. Don't pull the trigger. Squeeze it. Breathe out and squeeze."

The rifle cracked again. The bolt worked smoothly under Dick's hand. Down the range, the spotter called out a hit.

"Better," Harris said, and moved on.

Hundreds of rounds went downrange over the following weeks. Dick's shoulder developed a bruise that never faded. His grouping improved. He learned to account for wind, for distance, for the quirks of his rifle. Beside him, Tom struggled with the sighting, his shots consistently low.

"You're pulling down as you squeeze," Dick said quietly between volleys. "Try keeping your eye on the target, not the sight."

Tom tried it. His next shot hit closer to center. "Huh. That's better. Cheers."

By the end of their third week, Dick could fieldstrip his rifle in under a minute and group his shots within a hand's span at two hundred yards. He did what was asked without complaint, helped others when they struggled, and kept his equipment clean. Sergeant Harris noticed, though he said nothing directly. He gave Dick small responsibilities: leading a file during drill, checking equipment before inspection.

One afternoon during fieldcraft training, they were practicing section advances across open ground. Dick's section was to advance by pairs while others gave covering fire. The sun was directly overhead, the heat pressing down. They had been at it for over an hour, and Dick's throat was dry as paper.

Sergeant Harris called them up for another run. Pairing with Jack Morrison, Dick went forward at the crouch, rifle held ready, moving from cover to cover. Jack reached a small depression and dropped into it. Dick followed, landing hard beside him.

"Right," Jack said, breathing heavily. "Your turn to cover me."

Dick raised his rifle to firing position. Jack was about to move when Sergeant Harris's voice barked across the field.

"Roberts! What do you think you're doing?"

Dick looked up. Harris was striding toward them.

"Covering, Sergeant."

"Covering what? The sky? Your muzzle is pointed at the clouds. If Morrison moves, you're not covering anything." He pointed at Dick's rifle. "Lower. If you can't see what you're aiming at, you're useless."

Heat flooded Dick's face. He'd been so focused on getting into position that he hadn't checked his line of fire. He adjusted the rifle, bringing it down to a proper angle.

"Again," Harris said. "Both of you. Do it properly this time."

They ran the drill again. This time Dick kept his muzzle down, his line of fire clear. Jack moved forward to the next cover. Dick covered him properly, his sights tracking left and right for threats. When they reached the end of the advance, Harris gave a curt nod. "Better. Remember it."

That night in the hut, Jack was cleaning his rifle when he spoke. "Thanks for the tip on the range. About the sighting."

"You'd have worked it out."

"Maybe. But you got me there faster." He looked up. "And today, that business with Harris. Could've happened to any of us."

"It happened to me though."

"This time. Next time it'll be me or Tom or Bill making the cock-up." He went back to his rifle. "That's what we're here for. Working it out together."

The work required more than following orders. It required thinking, adjusting, and learning from errors. And having mates who didn't hold it against you when you got it wrong.

A month into training, Mick was assigned as a Bren gunner, with Bluey as his number two. The Bren was heavier than the rifle, fed by a top-mounted magazine that held thirty rounds. On the range, Mick settled behind the weapon, his large frame absorbing the recoil with ease. The Bren hammered out rounds in steady bursts. Bluey crouched beside him, feeding fresh magazines as soon as the first ran dry.

After one session, Mick came back with his shirt dark

with sweat and a grin on his face. "That's a proper weapon. You feel it working."

"You look like you've been wrestling it," Dick said.

"Near enough. But I reckon I'm winning."

In the evenings, after training was done and equipment cleaned, the men had a few hours to themselves. Letters home went out weekly. Training is going well. The sergeants know their business. Tell Dorrie that Bluey here talks more than all five of my sisters combined. Dick included bits that would make them laugh. Told them about the time Bill fell asleep during a lecture on map reading and started snoring loud enough to wake half the section.

The letters he received back came in multiple hands, Jean's neat script giving way to Lorna's looser writing, then Dorrie's careful print. They told him about home. About their father still working nights, their mother managing the household though it was harder without Dick there. About small dramas at school and church.

He wrote to Eileen every two weeks. Her replies came less frequently and said less each time. I'm glad you're well. Work is busy. The distance between them had grown wider than the miles separating Newcastle from Ingleburn.

Other men played cards. Frank was good at pontoon, his face giving nothing away. Bill was terrible at it but played anyway, losing his cigarettes with good humor. Tom had a decent voice and would sometimes lead them through old bush songs or music hall numbers.

One evening after a particularly hard day of field exercises, Dick sat outside the hut with Mick, both of them smoking, too tired for much conversation. Tom joined them, stretching out on the ground with a groan.

"Christ, my feet," Tom said. "I've got blisters on my blisters."

"That's nothing," Mick said. "Wait till we're humping full packs across the real country, not just training ground."

"Thanks for that cheerful thought."

They sat in silence for a while, watching the light fade across the camp. Then Tom spoke again, his voice quieter.

"You ever think about what it'll be like? When we get there. Wherever there is."

"Sometimes," Dick said.

"My dad," Tom said, "he was in France in the last one. He doesn't talk about it much. But sometimes I'd catch him just staring at nothing, and I'd know he was back there." He drew on his cigarette. "He told me once, the worst part wasn't the fighting. It was after, when you realized how many of your mates weren't coming back."

No one spoke for a moment. Then Mick said, "That's why we stick together. Look after each other. That's all we can do."

"Yeah," Tom said. "Reckon so."

As October turned to November, the training intensified. Forced marches with full packs, covering miles across rough country. Night training, learning to move in darkness, to navigate by compass and stars. Bayonet drills, charging at stuffed sacks while sergeants shouted encouragement.

Dick's body changed. His shoulders broadened. His hands became harder. The softness of civilian life wore away, replaced by lean muscle and hardened sinew. He could march twenty miles with a full pack and still be ready to dig in at the end. He could strip and reassemble his rifle in the dark. He could read a map, throw a grenade, move through scrub without making noise.

In mid-November, they were assembled on the parade ground at an unusual hour. Not dawn. Mid-morning. The whole camp, sections formed up in lines, waiting. Between Mick and Tom, Dick stood at attention, feeling the sun on his shoulders, the weight of his rifle.

The commanding officer appeared on the raised platform, his uniform crisp, his voice carrying across the assembled men.

"You are soldiers now," he said. "You have trained well. Your country needs you elsewhere. You will be moving out within the week. Further orders will follow. Dismissed."

That was it. No destination given. No explanation. Just the simple fact: they were going.

As the formation broke apart, Dick heard the murmur spread through the men. Egypt, some said. England. Palestine. Singapore. No one knew. But the preparations made it clear: overseas.

Inoculations came next, leaving their arms sore for days. Tropical uniforms were issued, lightweight and khaki. Packing and repacking their kit according to regulations, everything rolled and stowed in a specific order.

The night before they left Ingleburn, the men in Hut

7 sat up later than usual. No one suggested it. They simply were not ready to sleep. Bluey sat on his bunk pulling threads from a hole in his sock, for once not talking. Tom and Jack played a slow game of cards, neither of them concentrating. Bill lay on his back, hands behind his head, staring at the ceiling. Frank checked his kit for the third time that evening.

Mick sat on the edge of his bunk, cleaning his rifle again though it was already clean. Dick did the same, the familiar motions occupying his hands while his mind moved elsewhere.

"Reckon we'll stay together?" Frank asked finally. "When we get where we're going?"

"Should do," Tom said. "We're in the same section."

"We'll stay together," Mick said with certainty. "They don't split up sections. Not unless there's a reason."

"Right then," Bill said from his bunk. "That's settled."

But no one looked entirely convinced.

Eventually they climbed into their bunks. Dick lay awake for a while, listening to the sounds of the other men breathing, someone shifting position, the creak of a bedframe. Outside, the camp was quiet except for the wind moving through the gum trees.

They were soldiers now, or as close to soldiers as men could be without having seen war. The rifles they carried were familiar tools. The movements of drill had become automatic. Following orders, working together, enduring without complaint.

But they had not been shot at, had not seen men die, had not made decisions under fire. That knowledge waited for them across an ocean.

The next morning they assembled before dawn, packs on shoulders, rifles slung. Sergeant Harris walked the line one final time, checking straps and buckles, saying nothing. When he was satisfied, he nodded to the officer, who gave the command to march.

Dick fell into step with the others, his boots raising dust that caught the first light of the sun.

Trucks waited beyond the camp gates. Sections boarded in turn, Dick climbing up into the back of a canvas-covered Bedford with Mick, Tom, Jack, Bill, Frank, and the others from Hut 7. The truck's engine rumbled to life. The convoy moved out, heading toward Sydney, toward the harbour, toward the ship that would carry them across twelve thousand miles of ocean.

Near the back of the truck, Dick watched the countryside roll past through the open canvas. Paddocks and farmhouses and small towns where people stopped to watch them go by. Some waved. Some just looked.

The shipping office came to mind. His desk by the window where he could see the cranes moving at the docks. The house in Alfred Street. His sisters. His mother in the kitchen. His father with ink on his hands. Eileen standing under the awning of the pharmacy.

The images were clear and sharp, and he held them there for a moment before letting them go. There would be time to remember later. Now there was only the movement of the truck, the weight of the rifle across his knees, the presence of his mates around him, and the knowledge that the next part had begun.

CHAPTER 3 — THE VOYAGE

After six weeks waiting in camps outside Sydney, drilling and preparing while ships were assembled into convoy, the day finally came. The troopship Otranto pulled away from Sydney Harbour on a bright morning in January 1941. The wharf was packed with families who had come to see the men off. They pressed against the barriers, waving handkerchiefs and calling out names that grew fainter as the ship moved into open water. Women held children up to see. Old men stood with their hats in their hands. The sound of voices carried across the widening gap between ship and shore, a confusion of farewells that merged into one sound of leaving.

At the rail with Tom, Mick, and Bluey, Dick watched the city recede. Jack and Bill were somewhere further along, caught in the press of bodies. Frank stood a few feet away, his face pale. The ship's horn sounded, deep and final, and the deck vibrated beneath their feet. Below them, the water churned white as the propellers bit and drove the ship forward.

"There it goes," Mick said.

No one answered. Buildings grew smaller, people on

the wharf becoming indistinct. The harbour opened around them, wide and steel blue, with ferries crossing in the middle distance. Sydney had always been a name, a place other people went. Now Dick was leaving from it, and everything familiar was behind him.

Tom lit a cigarette and offered the pack around. They smoked and watched the Heads approach, those two cliffs that marked the entrance to the open sea.

"My mum's back there somewhere," Bluey said. "Stayed with my aunt in Newtown last night so she could see me off. Didn't see her in the crowd though."

"She was there," Tom said. "They're all there."

His own family in Newcastle had not come to Sydney. The trip would have cost too much. Dick had said his goodbyes at home, on the verandah, with his father's hand on his shoulder and his mother's face composed and steady. His sisters had waved from the front steps as he walked down Alfred Street with his kit bag.

The Heads passed on either side, massive and indifferent. Seabirds wheeled in the air, their cries sharp above the sound of the engines. The harbour fell away behind them. The ship rose and fell on the first swells of the open ocean, and the deck shifted under their feet.

The gulls followed for a time, diving and calling, then one by one they turned back toward land. The coast was still visible, a grey line on the horizon, already distant.

When it was time to go below, the space felt smaller than it had looked from the wharf. The Otranto carried over two thousand men. Hammocks hung in tight rows above tables and benches. The air was already heavy with the smell of bodies and oil and the salt-damp canvas of the hammocks. Each man had been assigned a number and a space no wider than his shoulders.

His hammock was in the third row, number 247. Dick stowed his kit beneath it. Around him, men were doing the same, their movements careful in the confined quarters. Someone bumped into him, apologized, and moved on. The noise was constant, a low roar of voices and footsteps and metal against metal.

That first night the sea was relatively calm, but by morning the Tasman showed its strength. The swells came from the south, long and deep, lifting the ship high and then dropping it again. Nothing stayed level.

By midday, men were sick. Some made it to the rails. Others did not. The smell below deck became worse, a mix of vomit and sweat and diesel fumes that seeped up from the engine room. The orderlies moved through with buckets and mops, but they could not keep pace. Men lay in their hammocks with their faces turned to the wall, groaning.

The motion rolled Dick's stomach but did not betray him. He had spent enough time on boats in Newcastle Harbour that the movement felt familiar. Not pleasant, but manageable.

Bluey was green-faced and miserable. He lay in his hammock gripping the sides, his usual stream of talk reduced to occasional groans.

"You going to live?" Dick asked.

"Don't know yet," Bluey said. "Ask me tomorrow."

"Do you need anything?"

"A bullet."

Water and a damp cloth for his face, that was all Dick could bring. There was not much else to be done. By the second day, some were adjusting. By the third day, most had found their sea legs or at least learned to function while feeling wretched.

Tom was sick once, violently, over the rail, and then seemed fine afterward. Mick was never sick at all. He moved through the ship as if the deck were solid ground.

"Grew up on a farm," he said when Dick asked. "Used to ride in the back of the truck on rough roads. This isn't much different."

The days developed a rhythm. Reveille at dawn, tea and hard biscuits for breakfast, then drill on deck if the weather allowed.

One morning in the Tasman, with the swells running high, Sergeant Harris assembled them on deck. The wind cut across the open space, cold and sharp. The ship rolled beneath them, tilting one way, pausing, tilting the other.

"Form up," Harris said.

They tried. The deck shifted and men stumbled, grabbing at each other for balance. Feet planted wide, Dick tried to anticipate the roll. Beside him, Frank lost his footing and went down hard on one knee.

"On your feet, Delaney."

Frank got up, his face red.

"Right face!"

Half the section turned right. The other half, confused by the motion, turned left. Two men collided, rifles clattering.

"Stop." Harris's voice was flat. "You think the enemy is going to wait while you get your balance? You think the ground stays still in combat?" He walked along the line. "Again."

They drilled for an hour, turning, marching, stopping, the deck rolling beneath them the whole time. It was harder than any drill at Ingleburn. Men fell. Rifles dropped. But gradually they learned to move with the ship, to anticipate the roll, to keep formation even when the world tilted.

By the end, aching legs and cramped hands from gripping his rifle. But Dick had stayed on his feet. They all had, more or less.

"Better," Harris said. "Dismissed."

When the weather was too rough for drill, they stayed below and cleaned equipment. The rifles had to be oiled constantly because of the salt air. Metal corroded fast at sea. They sat at the long tables, working oil into every part, checking the bolts and firing pins, reassembling everything with care.

In the afternoons they were allowed on deck for air. As much time there as possible—Dick preferred the wind to the crowded space below.

One afternoon, a week into the voyage, he leaned against a bulkhead watching four men play pontoon at a makeshift table. Bill Hughes was dealing, his movements practiced. Frank sat across from him, studying his cards. Two men Dick didn't know well rounded out the game.

"You in or out, Roberts?" Bill asked.

Content to watch, but the invitation pulled him in. "In."

He sat down. Bill dealt him two cards. A seven and a four. A moment's consideration, then a tap on the table for another. A nine. Twenty. He held.

Frank busted. One of the others made eighteen. Bill had nineteen and took the pot, a small pile of cigarettes.

"You're too lucky," Frank said.

"Not luck. Skill."

"It's all luck."

They played another hand. Dick won this one with twenty-one. Then Bill won. Then Frank, who grinned like he'd been proven right.

Tom appeared and leaned against the bulkhead. "Room for one more?"

"If you've got cigarettes."

He sat down. The game continued, the cards slapping down on the rough wood, cigarettes changing hands, small victories and losses that meant nothing and everything. It was a distraction from the endless water and heat.

"What do you reckon it'll be like?" Frank asked suddenly. "When we get there."

No one answered right away. Bill dealt another hand.

"Hot," Tom said finally.

"Hotter than this?"

"Probably."

"Christ."

They played in silence for a while. Then Frank spoke again, his voice quieter. "Are you a bit scared?"

"No," Bill said immediately.

But Tom looked at his cards and didn't answer. His hand folded, Dick stood up. "I'm out."

He went to the rail. The ocean was grey-green, the swells rolling in from the south. No land in sight, just water in all directions.

Tom joined him a few minutes later. They stood without speaking. Finally Tom said, "I didn't answer Frank's question."

"I noticed."

"Because the answer's yes. I am scared."

A nod. "Me too."

"Yeah?"

"Yeah."

Tom lit a cigarette. "I keep thinking about my wife getting a telegram. That's what scares me most. Not dying so much as her finding out."

"I think about my mum," Dick said. "And my sisters. How they'd manage."

"We could all die on this ship," Tom said. "Submarine could get us tonight. We'd never even see combat."

"That would be a waste."

"Would it though?" Tom drew on his cigarette. "Sometimes I think it'd be easier. No fighting, no having to shoot anyone, just gone."

"You don't mean that."

"No," Tom admitted. "I don't."

After a week at sea, they reached Fremantle. The convoy slowed and anchored in the roads outside the harbour. They were not allowed ashore. The stop was only for fuel and supplies. Lighters came out from the port, carrying crates and barrels that were winched aboard. Fresh water came too, and mail.

When the mail was distributed that evening, three letters arrived. His mother wrote about the weather, Meryl's lost tooth, his father's regards. Jean's was longer, full of news about neighbours and the beach. Eileen's was short, cordial and distant. She signed it "Yours, Eileen." Not "Love." Just "Yours."

The letters went into his kit, folded.

The next morning the convoy turned west into the Indian Ocean. Fremantle disappeared behind them. The next land they saw would be on another continent.

The heat intensified. The sun beat down on the metal deck until it was too hot to touch barefoot. Men who forgot yelped and danced. The air below deck became stifling, sweat soaking through shirts that never dried. At night the heat stayed, trapped in the metal hull.

Hands darker now from the sun, the skin tougher. His uniform hung a bit looser. Catching his reflection in a piece of polished metal one afternoon, Dick barely recognized himself. Face thinner, eyes harder. He looked older.

Tempers frayed. One morning a fight broke out over water rations. Two men grappled near the drinking fountain, shouting and swinging wildly. A corporal separated them with sharp words and the threat of punishment.

"It's the heat," Mick said. "Makes people stupid."

The heat made people quiet too. Even Bluey's constant commentary faded to occasional remarks.

Newcastle came to mind more often now. The beach. Diving into the surf on a hot day. The memory was vivid and useless.

One afternoon, as the ship plowed northwest, someone shouted and pointed. Flying fish broke the surface, a dozen or more, their bodies silver and sleek, their fins spread like wings. They skimmed above the waves for impossible distances before dropping back into the sea.

The men cheered. It was the first living thing they had seen in days besides each other.

"Did you see that?" Bluey said, animated again. "They were flying. Actually flying."

"Gliding," Mick corrected.

"Flying," Bluey insisted. "If it's in the air, it's flying."

They argued about it for the rest of the afternoon. Something to talk about, something other than heat and boredom.

Writing to Eileen proved impossible. The words felt false. What could he say? That he was well? That the sea was large? In the end, a short note thanking her for her letter and saying he was well. He signed it "Dick" and felt relief when it was done.

Three weeks into the voyage, the routine wore everyone thin. Men snapped at each other over small things. The close quarters, the heat, the endless sameness of ocean and sky pressed on them.

One evening, cleaning his rifle at a table below deck, Dick worked in a space crowded with men elbow to elbow. Bill was beside him, Frank across the table. Jack Morrison was further down, methodical as always with his work.

Someone bumped his arm, hard. The cleaning rod slipped and scratched across the barrel of his rifle.

He looked up. It was a man from another section, thick-shouldered, with a face red from the heat.

"Watch it," Dick said.

"Watch what?" The man's voice was aggressive. "You're the one taking up half the bloody table."

"I'm in my space."

"Your space." The man laughed, ugly. "There's no space. We're packed in here like bloody sardines."

"Then move carefully."

"Or what?"

He stood up. So did the man. They were the same height, but the other man was heavier. Around them, conversations stopped. Men turned to watch.

"Or nothing," Dick said. His heart was hammering but his voice stayed level. "Just watch where you're going."

"Are you telling me what to do?"

"I'm telling you to watch where you're going."

The man stepped closer. The smell of his sweat, the anger in his eyes.

Mick appeared between them. He didn't say anything,

just stood there, his size filling the narrow space. The red-faced man looked at Mick, then back at Dick.

"Not worth it," the man muttered, and turned away.

Sitting back down, Dick felt his hands shaking slightly. He picked up his cleaning rod and went back to work.

"All right?" Mick asked quietly.

"Yeah."

"Good." Mick moved back to his own space.

The rifle came clean with careful, deliberate movements. The incident was over. His hands kept shaking for ten minutes.

Sergeant Harris appeared at the end of the table, moving through on inspection. He paused at Dick's rifle, examined it, gave a brief nod. "Good work, Roberts." Then he moved on.

It was a small thing, but steadier after it.

At night, when he could, up on deck. The air was cooler then, and the stars were extraordinary. Away from the lights of cities and towns, the sky was dense with them. The Milky Way was so bright it cast shadows on the deck.

One night Tom found him there. They stood together in silence before Tom spoke.

"You ever wonder if we'll see home again?"

"Yes."

"Me too." Tom lit a cigarette, the match flaring briefly. "My wife cried when I left. I told her I'd be back. But I don't know that. None of us know that."

"No," Dick agreed. "We don't."

"I think about that sometimes. About not coming back." Tom drew on his cigarette, the tip glowing red. "Does that make me a coward?"

"No. It makes you honest."

They watched the stars in silence. Below them, the ocean was black and vast.

"Well," Tom said finally. "We'll find out soon enough."

After more than three weeks at sea, the convoy arrived at Colombo. The harbour appeared at dawn, shapes and colors emerging from grey light. The air grew warmer and heavier, humid and clinging. A scent came off the shore, unfamiliar, perhaps spices and vegetation and earth unlike Australia.

As the sun rose, details became clear. Ships filled the harbour, more than Dick had ever seen. Cranes lined the docks. Buildings rose behind the waterfront, white and pale yellow, with roofs of red tile. Palm trees lined the shore.

The men pressed against the rails, staring. This was the world beyond Australia, strange and immediate and utterly foreign.

Small boats appeared, clustering around the ship. Men held up fruit, carved wooden objects, necklaces made of shells. Some threw lines up to the deck. Soldiers hauled up baskets, lowered money, hauled the baskets up again. The sergeants shouted warnings but the trading continued.

In the afternoon, they were allowed ashore for a few hours. Tom, Mick, Jack, and Bluey went together. They walked along the dock, past warehouses and stacks of cargo, past buildings with signs in languages none of them could read.

The heat pressed down like something solid. Within minutes, their uniforms were soaked with sweat.

They found a market near the harbour. Stalls lined narrow streets, selling fruit and vegetables and fish and things Dick could not identify. The smells were intense, a mixture of spices and cooking food and human sweat and something sweet and rotten underneath it all.

"This is mad," Bluey said, turning in circles.

They bought postcards from a vendor who spoke a little English. Standing at a counter in a shop that sold tobacco, Dick wrote brief messages on six cards. To his mother: "In Ceylon now. Very hot. All well." To his sisters: short notes about the heat and the palm trees. To Eileen: "Still traveling. Hope you are well."

As they walked back to the ship, they passed a temple, its entrance guarded by stone figures. Smoke rose from inside, carrying a scent that was sweet and heavy. A man in orange robes sat near the entrance, his eyes closed.

Jack stopped, studying the building. "Look at the joints," he said quietly. "No nails. It's all fitted together. Must be hundreds of years old."

"How can you tell?" Bluey asked.

"The wear patterns. The way the stone's settled." Jack shook his head slowly. "Someone knew what they were doing."

They watched for a moment longer, then moved on. The

ship was visible ahead, grey and familiar.

That night, as the Otranto pulled away from Colombo and turned northwest, Dick stood at the rail and watched the lights of the city recede. The temple, the market, the boats, the man in orange robes. Nothing like Newcastle, any of it.

Three days out from Colombo, the convoy sailed on through the Arabian Sea, escorted now by destroyers. The threat of submarines was real. Ships had been sunk in these waters. The men were told to keep their life jackets close. Lifeboat drills were held every few days.

One night, an alarm sounded. Men scrambled from their hammocks, grabbing life jackets, rushing to their stations. Heart hammering, Dick climbed the ladder to the deck. Around him, men were pale and tense. They stood in the darkness, waiting, the ship moving beneath them, the water black below.

Nothing happened. After half an hour, the all-clear sounded.

"Bloody hell," Bluey muttered, climbing back into his hammock. "I thought that was it."

"So did I," Tom said.

Lying in his hammock, pulse gradually slowing, Dick thought about drowning, about going into that dark water. About his family receiving a telegram: Lost at sea.

He did not sleep well that night.

The days blurred as they continued north. The coast of Africa appeared to starboard, a brown line of desert and rock. Hostile and barren, nothing like the green coast of

Australia.

The Red Sea was narrow and hot, the ship moving between shores of sand and stone. The temperature climbed even higher. Men stayed below during the day, sprawled in their hammocks. The nights were better but not much.

One evening in the Red Sea heat, sitting on his footlocker below deck, Dick wrote a letter home. His shirt stuck to his back. Sweat dripped onto the paper, blurring the ink. Around him, men lay in their hammocks, too hot to move.

He wrote about the flying fish, about Colombo, about the heat. He did not write about the near-fight, or the submarine alarm, or Tom's fears. He kept his words light, factual. They would worry enough without knowing everything.

Finally, after six weeks at sea, the coast of Egypt appeared. It came as a pale band under the rising sun, flat and featureless. The convoy slowed. Ships began to separate, heading toward different ports. The Otranto moved toward Alexandria, past other vessels anchored in the roads, past patrol boats and minesweepers.

The harbour was vast and teeming with traffic. The smell reached them before they docked: diesel and salt and sewage and something dry and ancient that might have been the desert itself.

Orders were shouted. The men assembled with their kit, rifles slung, packs on shoulders. Forming into lines, they waited while the ship maneuvered to the dock.

When the gangway was lowered, they filed down onto the quay. Solid ground felt strange after weeks of movement. Legs adjusted slowly, expecting the deck to shift. Around them, the harbour was chaotic. Cranes swung cargo, trucks rumbled past, men shouted in Arabic and English.

Sergeant Harris moved along the line, checking that everyone was present.

"Welcome to Egypt," he said. "Form up and move out. The lorries are waiting."

Pack adjusted, Dick looked back at the Otranto one last time. Beyond the harbour, the sea stretched back toward Colombo, Fremantle, Sydney, Newcastle. All of it behind them now, impossibly far away.

"Roberts," Sergeant Harris said. "Move."

He turned and followed his section toward the waiting trucks. The voyage was done. Egypt surrounded them, hot and foreign and real. Somewhere beyond the city, beyond the desert, the war waited.

CHAPTER 4 — NORTH AFRICA

They had three hours in Alexandria. Three hours to climb down from the ship, form up on the quay, and load into the backs of trucks. Dick's legs kept adjusting for a deck that wasn't there. He stumbled on flat ground, overcorrected, nearly went to his knees. Six weeks at sea had rewired something in his balance.

The trucks were already running, diesel fumes mixing with the harbour smells. Sergeant Harris moved along the line, checking names, pointing men toward vehicles.

"Roberts, Morrison, Collins, Webb. That truck there. Move. "

Dick hauled himself up with Tom, Mick, and Bluey. Jack Morrison climbed in last, careful as always, and found a spot near the tailgate. Bill Hughes and Frank Delaney were being directed to a different truck, heading to a different section. Bill raised a hand in farewell. Dick nodded back. Ships, then sections. The army kept breaking them apart and putting them back together in new configurations.

Fifty men pressed together with kit and rifles, the

canvas cover doing nothing against the heat. The truck lurched forward. Through the gap in the canvas, Dick watched Alexandria fall away. White buildings, palm trees, the last glimpse of the harbour where the Otranto sat at anchor. Forty-two days getting here. Now they were leaving before the sweat from the voyage had dried.

The city gave way to scrub, then to sand, then to nothing at all. The road was barely a road, just two ruts worn into harder ground. The green of the Nile delta disappeared within an hour. After that, there was only brown and yellow and the occasional grey of rock. The sky was white with heat, the sun a disc of fire that seemed to fill half the horizon.

The heat came from above and below. The sun struck the metal frame of the truck while the sand reflected it upward again. Within twenty minutes, every man was soaked with sweat. Within an hour, the sweat had dried and left salt on their skin.

"Bloody hell," Bluey said, his voice hoarse. "And I thought the Red Sea was bad."

"This is worse," Tom said. His baby face sagged, mouth half-open, eyes unfocused.

Mick said nothing, just stared out at the desert with the same expression he'd worn watching the Indian Ocean. Like he was measuring something only he could see.

They stopped once for water and to let the trucks cool. The men climbed down stiffly. Dick's legs cramped from sitting, and the ground was hard and flat, covered with stones the color of old bone. No trees. No shade. Nothing but the trucks and the sun and the endless expanse of sand and rock.

"There's nothing here," Bluey said, turning in a slow circle.

"That's the idea," Mick said. "Can't hide an army in nothing."

They drank from their canteens, careful not to take too much. The sergeants had already started rationing water. A pint a day. The water was warm and tasted of metal and chemicals, but it was water.

Jack stood a little apart, studying the horizon the way he'd studied that temple in Colombo. "Strange place for a war," he said quietly.

"Strange place for anything," Tom said.

After fifteen minutes they climbed back into the trucks. The landscape never changed. Hour after hour, the same flat expanse, the same heat, the same white sky. Another planet. That's what it looked like. Forty-two days staring at water, now a thousand miles without a drop.

They reached camp in the late afternoon. It wasn't really a camp, just an area where other trucks were parked and tents had been pitched in rough rows. The ground was covered with the same stones as everywhere else, and the tents were the color of sand, barely visible against the landscape.

They set up their own tents quickly, working in pairs to drive stakes into ground that didn't want to hold them. Dick worked with Tom, neither of them speaking much. Six weeks on a ship together, and now this. The canvas smelled of dust and age. Inside, the heat turned the air to syrup. Breathing took effort.

"We're supposed to sleep in this?" Tom asked.

"Better than a hammock."

"Is it though?"

That first night proved Tom wrong, then right again. The heat stifled, but at least Dick could lie flat. He'd forgotten what that felt like. Still, sleep came hard. His body kept expecting the roll of the ship, and when it didn't come, something felt missing.

When dawn came and reveille sounded, the men dragged themselves from their tents, red-eyed and exhausted. Moving from ship to desert with no time between had wrung something out of them.

Breakfast was tea and hard biscuits and tinned beef that had gone grey in the heat. They ate standing up, swatting at flies. The flies were everywhere. More than on the ship, more than seemed possible. They landed on food, on lips, on eyes.

After breakfast came an inspection. Sergeant Harris walked down the line, his uniform somehow still neat. He stopped in front of Dick.

"Roberts."

"Sergeant."

Harris studied him for a moment. "Corporal Hayes bought it at Bardia last week. The section needs an NCO. Lieutenant says you"re it."

Hayes. Killed at Bardia. Dick tried to remember his face and couldn't.

"Yes, Sergeant."

"You've got eight men to start. Morrison, Collins, Webb, Patterson, the rest you'll meet. Keep them alive and keep them fighting. Understood?"

"Yes, Sergeant."

"Good." Harris moved to the next man.

Eight men. His. If they died, that was on him.

Tom stepped closer, his voice low. "Congratulations?"

"Something like that."

"You'll be good at it," Tom said. "You kept us sane on that bloody ship."

Had he? Dick wasn't sure. But he nodded.

The rifles needed attention immediately. Sand had gotten into everything during the truck ride. Dick gathered his section and showed them how to strip the weapons properly. Jack already knew, his movements were practiced and careful. Mick handled the Bren gun like an old friend. Tom fumbled with his bolt until Dick showed him the trick of easing it rather than forcing it.

"Like this." Dick moved over. "Ease it out, don't force it."

Tom tried again, and got it right. "Cheers."

This was what being corporal meant. Not just keeping his own kit clean, but making sure everyone else knew how as well.

Water was measured out twice a day from jerry cans that sat in the sun. Dick learned to drink before thirst made his hands shake, but not so much that his canteen was empty by midday. Small sips. Make each mouthful count.

The sun burned exposed skin within minutes. By

the third day, every man had learned to keep shirts buttoned despite the heat, sleeves rolled down, collars up.

At night the temperature dropped fast, and the stars came out extraordinary and vivid. After the cramped space below deck, after weeks of hammocks and bodies pressed together, Dick could lie on his back outside the tent and see the whole sky. The stars were so dense they cast shadows on the ground.

He wrote letters by torchlight. To his mother: *Made corporal. Nothing special, just means I'm responsible for keeping eight men and their rifles in good order. The desert is very hot but we're managing.*

To his sisters: shorter notes, nothing about the heat or the flies or how strange it felt to be on land that didn't move.

He still had that last letter from Eileen in his kit, the one he'd received in Fremantle. "Yours, Eileen." Not "Love." He'd written her that brief note from Colombo, and she'd probably received it by now, and there seemed to be nothing more to say. The distance between them was too great now, not just in miles but in everything else.

He didn't write to her again.

The change came on a morning in late February, ten days after they'd arrived. Dick woke to the sound of engines, distant but growing louder. Not the familiar rumble of trucks but something higher-pitched, more urgent.

He was out of his tent before he was fully awake, pulling on his boots, grabbing his rifle. Around him, other men were doing the same. He counted his section automatically. Seven present, one still in his tent.

"Aircraft!" someone shouted.

They came from the north, three of them, flying low and fast. Stukas. Dick had seen recognition charts on the ship but never the real thing. They had bent wings and fixed landing gear that hung below like claws. The sound of their engines rose to a scream as they dove.

"Take cover!" Sergeant Harris ran through the camp. "Get down!"

Dick threw himself flat behind a stack of supply crates. The screaming grew louder, mechanical and terrifying, a sound that went straight through the skull. Then the first bomb hit.

The explosion punched through his chest. His lungs forgot how to work. His stomach tried to climb into his throat. Dick pressed his face into the dirt, his hands over his head. Another explosion, closer. A third explosion, further away.

The scream of engines receded.

Dick lifted his head. Three columns of smoke. He counted them automatically, like counting his section. Men were shouting, running toward the impact sites. One of the supply trucks was on fire, flames pouring from its bed. The two tents nearby, flattened, canvas shredded.

"Wounded!" someone called. "We need medics!"

Dick got to his feet. His legs felt hollow, like the bones had gone soft. He counted his section. Tom was climbing out from under a truck, his face white. Mick moved toward the collapsed tents. Bluey stood frozen, staring at the burning truck. Jack knelt there, dazed but unhurt.

"Bluey!" Dick grabbed his arm. "Check on Patterson and the others. Make sure everyone's accounted for."

Bluey blinked, nodded, moved.

Dick followed Mick toward the collapsed tents. The canvas had been blown flat, and something dark was spreading beneath it. Mick lifted one corner and immediately dropped it again.

"Don't," he said to Dick. "Nothing to do for them."

But Dick had already seen. Two men, or what was left of them. He turned away and vomited into the sand, couldn't help it. His first dead men. They'd been sleeping.

"Happens the first time," Mick said. His voice was level, almost gentle. "Better to get it out."

Dick wiped his mouth, straightened up. "Right."

"Come on."

The raid had killed three men and wounded six. Dick knew two of them from the ship. Men who'd been in the hammocks near his, who'd played cards during the voyage, who'd stood at the rail watching flying fish. Now they were casualties. Medics loaded one into an ambulance truck, unconscious, his leg torn open.

They buried the dead that afternoon in shallow graves marked with crosses made from ration crates. The chaplain said words that were lost in the wind. Dick stood in formation with his section, watching Bluey carve names into the wood with his knife, the letters crude but careful.

After that, they dug slit trenches near their tents. Narrow holes just deep enough to lie flat in. When aircraft were spotted, they would drop in and wait, listening to the engines, trying to judge from the sound whether the planes were coming closer or moving away.

That afternoon, Dick was cleaning his rifle when Tom sat down beside him.

"Can I ask you something?"

"Yeah."

"That first raid. The bodies in the tent. You threw up."

"Yeah."

"But now you're just... you keep going. How?"

Dick thought about it. He thought about the ship, about the near-fight below deck, about the submarine alarm, about Tom saying he was scared. "I don't know that I do. There's what needs doing, and you do it. The rest, you think about later."

"And do you? Think about it later?"

"Sometimes." Dick looked at Tom. His hands were shaking slightly. "You all right?"

"I keep thinking, what if that'd been our tent?" Tom laughed, but it came out wrong. "What if next time it is?"

"Then it is." Dick stripped the bolt from his rifle, checked the mechanism. "But right now it isn't. So right now we keep our kit clean and we do what we're told. That's all there is."

Tom didn't look convinced, but he picked up his own

rifle and started cleaning it.

Raids came every few days. The Stukas would appear out of the white sky with that terrible scream, dropping their bombs, pulling up and disappearing. Dick learned to judge the sound. If the scream was getting louder, the plane was diving toward you. If it was steady or fading, the plane was diving somewhere else.

In early March they moved forward, driving northeast in convoy toward the coast. They passed burned-out tanks and abandoned vehicles and graves marked with helmets on rifles stuck into the sand. The war had been here already and left its marks.

They stopped for the night in a wadi. A dry riverbed that provided some shelter. Dick positioned his section along the northern lip, making sure they dug in properly.

"How deep?" one of the newer men asked.

"Deep enough that a shell lands five yards away, you're below the shrapnel," Dick said. "Two feet minimum."

He watched them dig. Eight men. His responsibility. Mick worked steadily, his trench already the deepest. Tom dug in fits and starts, pausing to wipe his face. Jack was methodical, measuring his trench like he'd measured that temple in Colombo. Bluey kept stopping to drink water until Dick told him to save it.

"You're in charge of water discipline now?" Bluey asked.

"I'm in charge of you not dying of thirst tomorrow. So yeah."

That night Dick took the second watch, two hours

of staring into darkness. The stars were brilliant. Somewhere out there were Germans and Italians, somewhere out there was the war. On the ship, it had seemed distant. Here, it felt close.

It found them the next morning.

Dust on the horizon, moving fast. Dick saw it first and called down to Harris.

"Enemy vehicles, Sergeant. Northwest, maybe three miles."

"How many?"

Dick squinted through the shimmer. "Three. Armoured cars."

The shout went up. Men scrambled to positions. Dick made sure his section was ready, checking rifles, positioning the Bren where Mick could use it properly.

The armoured cars moved south across their front, too far for effective rifle fire. Mick opened up with the Bren anyway, short controlled bursts. The cars didn't slow. Reconnaissance. They'd seen what they needed to see.

"They know we're here now," Harris said.

Two hours later, the shelling started.

The first round landed fifty yards ahead, throwing up sand and rock. Then more rounds, walking back toward their position.

"Down!" Dick pressed himself into his slit trench. The wadi ran east-west, their trenches along the northern lip, twenty yards from where the ground rose to a low ridge beyond. That's where the shelling was coming from.

The shells screamed as they came in, a rising whistle that ended in a crash. One landed close enough that Dick felt the shock through the ground. Sand rained down on him. His ears rang, a high whine that drowned out everything else.

"Roberts!" Harris's voice came through the ringing. "Your section!"

Dick lifted his head, checked. Tom was in the trench beside him, hands pressed over his ears but moving. Mick was already scanning for targets. Jack, Bluey, the others. All there, all alive.

"All present, Sergeant!"

The shelling stopped. Then the tanks came.

They appeared out of the shimmer, moving fast across the flat ground. Panzer IIIs, low and deadly. Behind them came infantry in halftracks, and behind them more tanks. Too many.

"Hold your fire," Harris said. His voice was steady, like he was back on the ship running drill. "Let them come closer."

Dick lay in his trench with his rifle aimed at the advancing line. His pulse hammered in his ears. The tanks grew larger. He could see the crosses on their turrets, the dust streaming from their tracks, the commanders in the hatches.

Three hundred yards. Two-fifty. Two hundred.

"Fire!"

The line erupted. Rifles cracked, the Bren hammered. The infantry took the worst of it, caught in the open. Men fell. Vehicles stopped.

But the tanks kept coming.

Dick fired at the infantry, worked the bolt, fired again. Beside him, Tom did the same, his face set and pale. Mick had the Bren singing, traversing across the enemy line.

One-fifty yards. One hundred.

The tanks" machine guns opened up. Dirt kicked up in front of Dick's face. He ducked down, came up, fired again.

A shell hit near Jack's position. The explosion was enormous. When the smoke cleared, Jack was on his side, not moving.

"Jack's down!" Tom shouted.

Dick wanted to go to him, but the tanks were at seventy yards now.

"Fall back!" Harris's voice cut through. "Withdraw to the trucks! By sections, covering fire!"

Dick grabbed Tom's shoulder. "Go! Mick, Bluey, get Jack!"

They pulled out, some covering while others retreated, then switching. Dick ran with Tom. His lungs burned, his pack heavy on his back. Behind them, the tanks ground forward.

Mick and Bluey dragged Jack between them. Blood streaked the sand.

They reached the trucks. Drivers had engines running. Men piled in, grabbing hands to pull others up. Dick counted his section as they scrambled aboard. Seven. Jack made eight, but Jack wasn't moving and his face was grey and blood ran from his ears and nose.

The trucks pulled away. One took a direct hit, exploding into flame. Dick looked away.

They drove south and west, the convoy scattered. Aircraft appeared. Hurricanes that dove on the German tanks. But it was temporary. A reprieve, nothing more.

They stopped at nightfall. Jack died in the back of the truck before they could get him to the medical tent. Dick, Tom and Mick lifted him down and laid him in the sand. His eyes were half open. Dick reached out and closed them. His hand shook now that there was time to shake.

"He was careful," Mick said. "Always careful with everything. Didn"t matter."

Dick thought about Jack on the ship, Jack studying that temple in Colombo, Jack asking about stone and joints and how old things were built. Two days ago he'd been cleaning his rifle, methodical as always. Now he was dead because a shell hit close enough and being careful didn't matter.

They sat with their backs against the truck wheels, drinking water, not talking. Tom's hands trembled as he lit a cigarette. Mick cleaned the Bren gun even though it didn't need cleaning. Bluey stared at nothing.

"Now we know," Bluey said finally.

"Know what?" Tom asked.

"What it's like."

No one answered. They all knew now.

Dick didn't write letters that night. He lay in his bedroll under the stars and tried not to think about Jack's half-open eyes, about how random it all was. Exhaustion pulled him under.

The days after fell into a pattern of movement and waiting. They dug in, moved out, dug in somewhere else. The heat never let up. The flies never left.

One afternoon outside Tobruk, Dick noticed Tom hadn't touched his water ration.

"You need to drink."

"Not thirsty."

"Drink anyway. That's an order."

Tom looked at him with something blank in his eyes. Like he was watching from somewhere far away.

"You all right?"

Tom leaned closer. "Jack keeps talking to me."

"Tom—"

"I know he's dead. I'm not mental. But I keep hearing his voice."

Dick didn't know what to say to that. "You need the medic?"

"No." Tom took the canteen, drank. "I just need to stop thinking."

"Then don't. Think about after. Think about home."

"Can't remember what it looks like anymore. Home. Can you?"

Dick tried to picture Newcastle, his house, the ocean. The images came but felt like they belonged to someone else. "Not really."

"Didn't think so."

That night, Dick pulled Mick aside. "Keep an eye on Tom."

"None of us are right," Mick said. "But I'll watch him."

"Thanks."

"You're doing all right, you know. As a corporal."

Dick looked at him, surprised.

"You're keeping us moving, keeping us focused," Mick said. "That's all anyone can do. Hayes tried to keep everyone safe and got himself killed. You just keep us doing the next thing. That's better."

Dick nodded, though he didn't feel like he was doing anything right.

The mail came through in late March. Dick's mother wrote that they'd harvested the tomatoes. Jean wrote about the hospital. Dorrie wrote about school. The letters felt like they came from another world.

He wrote back: *Made corporal. Nothing special, just means I'm responsible for eight men. Seven now. One of ours didn't make it. The desert is hot but we're managing.*

He didn't write about Jack dying in the truck. He didn't write about Tom hearing voices. He didn't write about how his hands still shook sometimes when the shelling started. His family didn't need to know.

In early April, new orders came. They were moving, but not forward. Back to Egypt to regroup. But some units were being sent somewhere else.

Dick's unit was going somewhere else.

The rumours said Greece. The rumours said Crete. The rumours said they were being sent to face a German invasion that was already underway.

They packed their kit and boarded ships in Alexandria. Dick stood at the rail and watched Africa recede. They'd survived the desert. Most of them. His section had lost Jack but the rest were still alive. Tom was fragile, Bluey was drinking when he could get it, but they were managing. Mick was steady.

Dick had learned to function in combat. His body knew what to do when the firing started. Fear was always there, but it was background noise now. What mattered was the immediate. The rifle in his hands, the men beside him, the next order.

He'd learned that men died and you kept moving because there was no time to stop. He'd learned that being corporal meant making decisions that might get people killed, and making the next decision anyway. He'd learned that you could watch a friend die in the morning and eat your ration in the afternoon because not eating wouldn't bring him back.

The desert had marked them. Dick looked at his reflection in the ship's dark window and barely recognised himself. His face was dark and creased. His eyes had changed.

The ship turned north into the Mediterranean. Six weeks ago they'd crossed this sea heading the other way, exhausted from the voyage out, nervous about what waited. Now they knew. Now they were veterans of a sort, though Dick didn't feel like one.

He thought about his father teaching him to swim in

the ocean at Newcastle. The memory was clear but felt like it belonged to someone else. Some younger version of himself who'd never seen a man die.

Ahead lay more war. Greece, Crete, wherever they were going. The Germans would be there first. They always were.

Dick touched the corporal stripes on his sleeve. Seven men now. Jack's spot still felt empty. Tom wasn't right. But they were his, and he'd keep them alive as long as he could.

The Mediterranean stretched away in all directions, blue and calm, nothing like the desert. Somewhere ahead was the next battle, the next retreat, the next death.

Dick turned from the rail and went below to find his section. Seven men. Jack's spot still empty.

CHAPTER 5 — GREECE

The convoy left Alexandria on a grey morning in early April. Dick had found his section below deck the night before. Just seven men in their hammocks, Jack's space empty among them. He counted them one more time before coming back topside. Now he stood at the rail with Tom, watching Egypt recede into the haze. Behind them lay the desert, the endless flat waste they had fought across and retreated from. Ahead lay something different, though no one in the ranks knew exactly what.

"Greece," Tom said. He said it carefully, like the word might break.

"Never thought I'd see Egypt either," Dick replied. "Or any of this."

The swells were long and gentle, easier than the Indian Ocean had been. After the desert's heat, the sea air was cool, almost cold. Men who'd baked under the African sun now pulled on extra shirts and huddled in whatever shelter they could find on deck.

Below deck, the space was crowded and close, hammocks strung in the same tight rows as on the voyage from Australia. But the men were different now. Thinner, harder, darker from the sun. Some needed help getting out of their hammocks. Others moved slowly, like old men. They had seen combat. They had seen friends die.

Away from the press of bodies, Dick found a spot on deck and sat with his back against a ventilator housing. He pulled out paper and a pencil from his pack. He had been putting off writing home, not from lack of desire but from not knowing what to say. How do you tell your family about the desert? About the heat and the flies and the retreat? About shooting at men and watching them fall?

But Lorna deserved honesty. She always had. Of all his sisters, she was the one who would see through careful words to what lay beneath. The third oldest but the most vocal, the one who argued with Jean about everything, who had opinions about things she wasn't supposed to have opinions about. She would be seventeen now. Nearly grown. And she would want the truth.

He began to write.

Dear Lorna,

I'm on a ship again, heading north. They say Greece this time, though no one tells us much. We just go where we're told and do what we're ordered. That's how the army works, I suppose.

I'm writing to you specifically because I think you can handle hearing how things really are. The desert was hard. Harder than I expected. We fought and then we pulled back, and then we fought again and pulled back again. That was the pattern. Men died. Good men, men I knew. I won't give you their names because I don't want you mentioning them to their families if you know them. That's not my place.

The heat was worse than any summer in Newcastle. Water was rationed to a pint a day. The flies were everywhere, in your food, your water, your eyes and mouth. At night it got so cold you'd shiver. Nothing about it was like what I imagined war would be.

I think about home more than I probably should. I think about the beach and the sound of the surf at night and Mum's cooking and all of you girls arguing around the kitchen table. I can picture it so clearly sometimes that it feels like if I just turned around, I'd be there. But then I open my eyes and I'm on a ship in the middle of the Mediterranean, heading to another country to fight in another battle.

How are you managing at home? I know it must be hard with me gone and Dad working nights. You and Jean must be carrying more of the load. I hope you're not fighting too much. Or maybe I hope you are, because that would mean things are still normal there, that the war hasn't changed everything.

Tell me honestly in your next letter, how is Mum really doing? She writes that everything is fine, but I know her. She's probably worried sick and not saying it. And the younger ones, Dorrie and Shirley and Meryl, are they understanding what's happening? Or are they too young to really grasp it?

I saw a man die last week. I won't tell you how or why, but I'll tell you this: it happened fast, and there was nothing anyone could do. One moment he was there, talking about his wife and his farm in Queensland. The next moment he was gone. That's what war is, Lorna. Not the glory and the flags and the speeches. Just men dying fast, and the rest of us carrying on because there's nothing else to do.

I'm telling you this because you're old enough to hear it, and because someone at home should know what it's really like out here. Not the censored version in the newspapers. Not the letters that say everything is fine and we're doing our duty. The real version.

We're heading to Greece, and from what the officers say, it won't be easy. The Germans are pushing south through the Balkans. We're supposed to help the Greeks hold them back. I don't know if we can. We couldn't hold them in the desert, and there were a lot more of us there. But we'll try. That's all we can do.

Keep your chin up, Lorna. Look after Mum and the little ones. Tell Jean she's right about whatever you're arguing about, even if she isn't. It'll surprise her so much she might actually stop for a minute. Tell Dorrie to keep up with her schoolwork. Tell Shirley and Meryl I miss them.

And don't worry about me. I'm still here, still whole, still doing what I came here to do. I'll write again when I can.

Your brother, Dick

He read it over once, then folded it and put it in an envelope. The censor would read it before it went out, might even cut parts of it. But most of it would get through. Lorna would read it and understand what he was trying to say.

Tom appeared and sat down beside him. "Writing home?"

"Just finished."

"How do you tell them what it's like?"

"You don't," Dick said. "You can't. But you try anyway."

They sat in silence for a while, watching the sea pass. Other men moved around the deck, some talking quietly, some sitting alone. Men spoke in low voices or didn't speak at all. No one laughed. Everyone knew they were heading into another fight, and everyone remembered how the last one had ended.

"You scared?" Tom asked.

Dick thought about it. "Not scared exactly. More like... resigned. We'll fight, and some of us will die, and the rest will do what we're told until the next time. I don't see how it ends any other way."

"Cheerful thought."

"You asked."

Tom pulled out a cigarette and lit it, cupping his hands around the match against the wind. He took a drag and exhaled slowly. "My wife's pregnant. Got a letter just before we left Egypt. Due in September."

Dick looked at him. "Congratulations."

"Thanks." Tom stared out at the water. "Means I have to stay alive, doesn't it? Can't leave her to raise a kid alone."

"Then stay alive."

"That's the plan." Tom laughed without humour. "Not

that plans mean much anymore."

On the third day, aircraft appeared. High at first, just silver specks against the sky. Then diving, engines screaming. Stukas.

"Take cover!" someone shouted.

Dick threw himself flat against the deck. Around him, men scrambled for the hatches. Too many bodies, too narrow an opening. Men jammed in the doorway, shouting, pushing. Dick pressed himself against the ventilator housing and watched the Stukas dive.

The first bomb hit water fifty yards to starboard. A column of white spray shot up, hung suspended, crashed back down. The ship heeled hard to port, the deck tilting beneath him. Dick grabbed a stanchion to keep from sliding.

The second bomb hit the transport three ships over. The explosion was orange and black, the shockwave visible as a ripple in the air. The ship's bow lifted, metal shrieking. Men poured from the deck like ants from a kicked nest.

The convoy's anti-aircraft guns opened up, filling the sky with black puffs of smoke. The Stukas pulled out of their dives and circled for another pass. Then fighters appeared. Hurricanes, diving on the German bombers, driving them east.

The damaged transport listed badly now, men jumping into the water. Destroyers moved in to pick up survivors while the rest of the convoy kept moving. There was no time to stop. The sea swallowed some. Others were pulled from the water, oil-slicked and coughing.

Dick got to his feet. He made fists to stop his hands trembling. Tom crawled out from under a lifeboat. The color had gone out of his face.

"You all right?" Dick asked.

"No." Tom looked at the sinking ship. "But I'm alive."

The Greek coast appeared the next morning. Mountains climbed straight from the sea, covered in green so bright it hurt to look at after months of desert brown. Small white villages clung to the hillsides. The water near shore was turquoise. Dick could see the bottom twenty feet down. After the desert's dust-choked air, the clarity felt strange.

They disembarked at Piraeus. Dick marched down the gangway into chaos. Supplies piled on docks, troops heading inland, refugees trying to board ships going south. The air smelled of pine resin and something sharp and green. Not dust. Not diesel. Dick couldn't name it but his lungs didn't burn with it.

Greek civilians watched them pass through the streets. Some waved. Some just stared. Children ran alongside asking for chocolate or cigarettes. Old women dressed in black stood in doorways with unreadable expressions.

The trucks drove north into the mountains. The roads were narrow and winding, climbing through forests of pine and oak. Green everywhere. Too much of it. Dick's eyes couldn't adjust to so much color after months of desert brown. Streams ran clear beside the roads. Wildflowers grew in the meadows.

They stopped on a ridgeline overlooking a valley.

Below, a road ran east through olive groves toward the coast. Beyond the road, the valley floor rose gradually for three hundred yards before climbing steeply to a wooded ridge on the far side. That's where the Germans would come from, the lieutenant said. From the north, pushing south.

"Dig in," Sergeant Harris said. "Two feet deep minimum. Fields of fire overlapping. Bren gun on the left flank where it can cover the road."

The section spread along the ridge. Seven men with overlapping fields of fire. Tom on Dick's right, Mick with the Bren gun twenty yards to the left, Bluey feeding ammunition. The others spaced between. The soil was dark and rich, easy to dig. After the desert's rock and sand, the trenches went down fast.

"How long do you reckon we'll hold here?" Bluey asked, shovelling dirt.

"As long as we're told to," Dick said.

"That's not an answer."

"It's the only one I've got."

Three days to prepare. Three days to dig and site weapons and watch the northern ridge. Three days of waiting while Greek refugees streamed south along the road below, carts piled with belongings, old people walking, children crying.

On the morning of the fourth day, the Germans came.

German artillery opened up first. Shells screamed in from the north and exploded along the ridge. Dick pressed himself into his trench. The ground shook. Dirt rained down. His ears rang with the familiar high whine.

Then the Stukas came, diving and bombing. Then Messerschmitts, strafing the positions with machine gun fire. The anti-aircraft guns fired back but there weren't enough of them.

When the barrage lifted, Dick raised his head and looked across the valley. Movement in the trees on the far ridge. Grey uniforms appearing, disappearing. German mountain troops, advancing in rushes, using cover.

"Here they come," Harris shouted down the line. "Wait for the order."

The Germans came down the far slope in sections, one group moving while another provided covering fire. They were good. Better than the Italians in the desert. More disciplined, more aggressive.

They reached the valley floor and started across. Four hundred yards. Three-fifty. Three hundred.

"Fire!"

The rifle kicked against Dick's shoulder. A German fell. Dick worked the bolt, found another target, fired again. Beside him, Tom fired steadily, his face set and pale. Down the line, Mick's Bren gun hammered, traversing across the advancing troops.

The Germans took cover in the olive groves. Mortar rounds started falling on the ridge, dropping with terrifying accuracy. One landed ten yards from Dick's trench, throwing up dirt and rock. His ears rang worse than before.

The Germans pushed forward again. Two hundred

yards now. Dick fired until his rifle was hot, the barrel singing. Empty magazines piled beside him. Tom passed him a fresh one.

"They just keep coming," Tom said. His voice was tight.

"Then we keep shooting."

The Germans reached a stone wall at the edge of the olive grove. They were setting up machine guns there, Dick realised. Once those guns opened up, the ridge would be untenable.

"Mick!" Dick shouted. "Stone wall, two o'clock! Machine gun crew!"

The Bren gun swung over. Mick fired a long burst. The Germans at the wall scattered. One didn't get up.

The battle went on for hours. The Germans would advance, take casualties, and fall back. Then the mortars would pound the ridge. Then the Germans would advance again. Dick fired until his shoulder ached from the rifle's recoil, until his hands were black with powder residue, until he couldn't remember how many magazines he'd gone through.

Then word came down the line. New German units were moving around their right flank. They had to pull back or be surrounded.

"Withdraw!" Harris moved along the position. "By sections. First section covers, second section falls back fifty yards, then reverse. Move!"

Dick grabbed his rifle and pack. "Tom, Bluey, go! Mick, you're with me. We'll cover."

Tom and Bluey scrambled out of their trenches and ran back into the trees. Dick and Mick fired down into the valley, keeping the Germans" heads down. When Tom's group stopped and took up firing positions, Dick tapped Mick's shoulder.

"Our turn. Go!"

They ran through the trees, bent low, lungs burning. German bullets cracked past. Dick dove behind a fallen log, set up his rifle, and started firing. Mick set up the Bren beside him.

"Bloody shambles," Mick muttered, feeding a fresh magazine into the gun.

"That's war," Dick said.

They fell back through the afternoon, section by section, position by position. The Germans followed, pressing hard. By dusk they'd given up two miles of ground and the Germans showed no sign of stopping.

They walked through the night. No time to rest, no time to eat. Just walk, stumble, keep moving south. German flares went up behind them, lighting the sky. Artillery fell randomly, probing for them in the darkness.

Tom stumbled and nearly fell. Dick caught his arm.

"I'm all right," Tom said.

"No, you're not. None of us are. But keep walking anyway."

"That baby," Tom said. "September. Got to make it to September."

"You will."

"You don't know that."

"No," Dick agreed. "But we'll act like I do."

Tom nodded and said nothing more. They kept walking through the darkness, the column of exhausted men stretching ahead and behind.

The pattern repeated. Another ridgeline, another defense. The Germans attacked, the Australians held for a few hours, then pulled back before being flanked. The Greeks were falling back too, and the British units. The whole line was collapsing.

On the fifth day they reached a village. Stone houses lined a narrow street, civilians hiding in cellars while bullets cracked overhead. Dick's section held the eastern edge, firing from windows and doorways while the Germans advanced through the olive groves beyond.

Mick had the Bren gun set up in a second-storey window. He worked the gun with his usual precision, firing controlled bursts, changing barrels when one got too hot. Bluey crouched beside him, feeding ammunition. Dick noticed the smell of alcohol on Bluey, faint but present. Not enough to matter right now. Later, maybe, Dick would say something. But not now.

"How much have we left?" Mick asked.

"Four magazines," Bluey said.

"Christ."

In the house next door, Dick watched through the window. Germans moved through the trees, getting closer. His ammunition was running low too. They all were.

A mortar round hit the roof above them. The air cracked. Part of the ceiling came down. Dick shook off the dust and debris, checked himself for wounds. Nothing. Tom had a cut on his forehead, blood running down into his eye, but he was still moving.

"We need to get out of here," Tom said.

"Not yet. Wait for the order."

The order came five minutes later. Fall back to the next village, two miles south. Dick ran across to Mick's position.

"Time to go. Bring the Bren."

They ran through the village, bullets chasing them. An old woman stood in a doorway, watching them flee. Her face showed nothing.

At the southern edge they regrouped. Dick counted his section. Five present. Two missing.

"Where's Patterson and Webb?" he asked.

No one knew. Dick waited two minutes, scanning the village. No sign of them. Dead or captured or lost. No time to find out which.

"Move out," Harris said.

Dick turned and followed. Five men now. Down from seven, down from eight.

On the eighth day they reached the coast. The small port was packed with troops, all trying to get aboard the few ships available. German aircraft circled overhead, diving to bomb whenever they saw targets.

Harris assembled what remained of the platoon in an olive grove near the beach. Thirty men, maybe. Half what they'd started with.

"Embarkation's tonight," Harris said. "After dark. Get some rest if you can. No fires, no unnecessary movement. Germans have artillery in range."

Dick found a tree and sat with his back against it. His feet were raw from walking. His shoulder was bruised from rifle recoil. His hands were cramped from gripping his weapon for so many days. Tom sat down beside him. Neither spoke.

The day crawled past. Men tried to sleep but few managed. The waiting was worse than the fighting. At least in combat there was something to do, someone to shoot at. Here they just sat and waited and hoped the Germans wouldn't shell them before darkness came.

Around midday, artillery started falling. The first rounds hit the beach, throwing up sand and spray. Then they walked inland, searching for targets. One landed in the olive grove fifty yards away. Dick's ears popped from the pressure. Shrapnel whined through the trees, cutting branches that fell with soft thuds.

"Anyone hit?" Harris shouted.

Voices called back. One man wounded, leg torn up. The medics moved in.

More shells fell. Dick pressed himself flat beneath the tree. The ground shook. Olive branches rained down, cut through by shrapnel. A shell landed closer, twenty yards. Dick felt the concussion through the earth.

Then it stopped. The silence after was almost worse than the noise.

Dick lifted his head. Smoke hung in the grove, thick and white, smelling of cordite and torn olive wood. A man screamed. Other men moved, checked themselves, called names.

"You all right?" Dick asked Tom.

"Still here."

"Mick? Bluey?"

"Here," Mick's voice came back.

Bluey didn't answer. Dick crawled over to where he'd been. Bluey lay on his side, his red hair dark with blood. Shrapnel had caught him across the back and shoulder. He still breathed but didn't respond.

"Medic!" Dick shouted.

A medic came running, took one look, and started working. Dick moved back. There was nothing he could do. Either Bluey would make it or he wouldn't.

The medic looked up after a few minutes. "He needs to get on a ship tonight. He won't last otherwise."

Dick nodded. Four men left in his section.

The afternoon dragged on. More aircraft came, more bombs fell. More men died within sight of the ships that were supposed to save them.

When darkness finally came, Harris got them moving. "Stay quiet. Stay together. Move when I tell you to move."

They moved Bluey first. Two men carried him on a stretcher down toward the water. Dick walked beside them. Bluey had come conscious, his face grey with pain.

"Fucking shrapnel," Bluey said. His voice was weak but he was trying to grin. "Never even saw it coming."

"You'll be all right," Dick said.

"Better be. Got things to complain about back home." Bluey gripped Dick's arm for a moment. "Keep them safe, yeah?"

"I'll try."

They reached the water's edge. A boat was waiting, already crowded with wounded. They loaded Bluey aboard carefully. The boat pulled away toward the dark shape of a destroyer offshore. Dick saw Bluey's red hair for a moment in the darkness, then the boat was swallowed by the night.

Three men left now in his section.

Dick returned to the olive grove. The rest of his section moved out in the next group, crossing open ground to reach the beach. German flares went up periodically, lighting the night. When a flare went up, everyone froze. When it died, they moved again.

Dick reached the beach and waded into the water. Cold, shockingly cold after months of heat. The water came up to his waist, then his chest. Another boat appeared out of the darkness, already crowded with men.

"Room for four more," someone said.

Dick, Tom, Mick and one other man climbed aboard. The boat sat low in the water, dangerously overloaded. The engine coughed and started. They moved away from shore.

Behind them, fires burned where the Germans had bombed. Ahead, the dark bulk of a destroyer loomed. Dick looked back at Greece one last time. They'd fought there for eight days. They'd lost.

The boat reached the destroyer. A rope net hung down the side. Dick grabbed it and started climbing. His arms shook from exhaustion. Hand over hand, foot by foot. He nearly fell twice but kept going. At the top, hands reached down and pulled him aboard.

He collapsed on the deck. Around him, other men did the same. Soldiers packed the deck, barely room to sit.

Tom climbed over the rail and fell beside him. Mick came next, still carrying the Bren gun. More men climbed aboard. The deck filled.

The destroyer waited until it couldn't hold any more, then pulled away from shore. More men were left behind. Dick knew. Thousands of them. Dead or captured or waiting for ships that wouldn't come.

He sat with his back against a bulkhead and stared at nothing. Egypt, Greece. Two defeats. Two retreats. The pattern was clear now.

Tom found him and sat down, saying nothing. What was there to say? They were alive. That was more than many could claim.

The destroyer moved through the darkness, engines thrumming. Around them, men talked in low voices. Crete, someone said. That's where they were going. Another island to defend. Someone else had heard the Germans were trying something different this time. Airborne assault. Paratroopers dropping by the thousand.

Dick closed his eyes. He was tired in a way sleep couldn't fix. Tired from the inside out, like something essential had been ground away. He'd been away from home for over a year now. He'd crossed oceans and continents.

He'd fought in two countries and lost both times. He'd watched friends die.

And he was still here, still alive, still moving toward the next battle.

He didn't know how much longer he could keep doing this.

But he would do it anyway, because that's what soldiers did. They kept going until they couldn't anymore.

Crete. Another island. Another airfield to hold, probably. Another battle waiting.

He'd heard enough. He'd see for himself soon enough.

One battle at a time. That's all he could manage now.

The destroyer sailed on through the night, carrying them toward whatever came next.

CHAPTER 6 — MALEME

The destroyer reached Suda Bay at dawn. Dick stood at the rail watching Crete take shape through the morning haze. Mountains rising straight from the sea, brown and barren in a way that reminded him too much of the desert. No green here. No forests like Greece. Just rock and dust and olive trees clinging to the slopes.

Tom appeared beside him. Neither spoke. What was there to say? Another island. Another position to hold until they couldn't anymore.

The section assembled on deck before disembarking. Three of them from the original eight: Dick, Tom, and Mick with his Bren gun. Four replacements from other units, men whose names Dick barely knew. Walsh, a quiet corporal from another section. Harris, red-haired like Bluey but younger, barely twenty. Two others Dick would need to learn as they went.

Sergeant Harris gathered them before they went ashore. They called him that to distinguish him from Young Harris, the red-haired replacement.

"Maleme airfield. That's where we're going. You'll dig in and wait. Intel says the Germans are planning something different this time. Airborne assault. Paratroopers."

"How do you land troops from the air?" Walsh asked.

"That's what we're going to find out." Harris looked at each of them. "This position matters. We hold the airfield, we hold the island. Clear?"

They nodded. Dick had heard the same speech in Egypt, in Greece. This position matters. We have to hold. Then the Germans came and they pulled back and it started again somewhere else.

The trucks ground west along the coastal road through the afternoon. Dick dozed in the back, his rifle across his lap. The exhaustion from Greece sat heavy in his bones. He'd slept on the destroyer but it hadn't been enough. Nothing was ever enough anymore.

They reached Maleme late in the day. The airfield sat in a shallow valley between hills and sea. A flat stretch of packed earth with a single runway running east to west. Olive trees on the slopes above. The sea is close enough to smell.

Tom climbed down from the truck beside Dick.

"Not another bloody airfield," Tom said.

"At least this one's got trees," Mick replied.

Dick looked at the slopes, the valley, the single runway. Nowhere to retreat to this time. Water behind them, mountains ahead. They'd hold here or they'd lose everything.

They drew positions on the north slope where the ground rose toward the hills. A New Zealand battalion held the airfield itself. The Australians would cover the approach from the sea and support the Kiwis if needed.

Harris pointed to a line of olive trees halfway up the slope. "Dig in there. You can see the whole runway from those trees."

The ground rang when the shovels struck. Rock and limestone, hard as the desert. Dick's hands had calloused over months of digging but this was harder still. Every strike threw sparks, every shovelful fought back. They managed shallow scrapes rather than proper trenches.

Walsh worked beside Dick, both grunting with effort.

"Better than the desert at least," Walsh said. "We've got shade and water."

"We've got nowhere to go," Dick replied. "That's what we've got."

Walsh stopped digging. "You think we can't hold?"

"I think everywhere we go, we end up leaving." Dick went back to his shovel. "I'd rather be somewhere we can stay."

Walsh didn't answer. None of them did. They all felt it. The pattern of arriving, digging, fighting, retreating. Egypt, Greece, now Crete. Each time the brass said this position mattered. Each time they'd been pushed back.

By evening they'd scraped out enough shelter to lie flat. Dick sat among the olive trees as the sun dropped toward the sea. Below, the airfield sat quiet. A few aircraft at the eastern end, covered with camouflage netting. Trucks moving along the perimeter. Everything is ordinary, peaceful.

Young Harris sat against an olive tree cleaning his rifle

for the third time. They were calling him that now to keep him straight from the sergeant.

"You reckon they'll really come from the air?" he asked.

"That's what they're saying." Mick checked the Bren, making sure no dust had fouled the mechanism.

"But how?" Young Harris looked up. "How do you get infantry out of an aircraft?"

"Gliders maybe," Tom said. "Or they drop supplies first, then land on the airfield once it's clear."

"Either way," Dick said, "they have to come down. And when they do, we shoot them."

But none of them sounded convinced. Armies fought on the ground. Aircraft supported from above. That's how war worked. This new German idea, whatever it was, fit no pattern they understood.

Sergeant Harris came through at dusk, checking fields of fire. "When they come, wait until they're close. Don't waste ammunition. Bren teams take aircraft if they try to land. Riflemen take whatever presents itself."

"Yes, Sergeant."

Harris looked at each of them, his face shadowed. "This airfield matters. We hold it, we hold Crete. We lose it, we lose the island. Understood?"

"Yes, Sergeant."

"One more thing." Harris paused. "We don't know what they're planning. No one's tried this before. So stay alert and wait for orders. Could be supplies, could be troops, could be something new. Just be ready."

After he left, they sat in silence. The not knowing was

worse than combat. At least in the desert you knew what to expect. Tanks, artillery, infantry in lines. Here they were waiting for something none of them could imagine.

Dick lay in his shallow trench that night and looked at the stars. Same stars as Newcastle, as the desert, as Greece. Constant. He thought about home. About the beach, about his mother's cooking, about his sisters crowded around the kitchen table. He'd written to Lorna on the ship from Egypt. Told her the truth about the desert, about the retreat, about how tired he was. She'd understand. She always did.

A sound brought him back. Aircraft engines, faint and distant. Others heard it too. Men sat up, listening. The sound came from the north, multiple engines, growing neither louder nor softer. Then it faded.

"Reconnaissance," someone whispered.

Dick lay back down. His rifle was beside him, loaded. His canteen is full. Sixty rounds in his pouches. Everything he needed to do his job. All that remained was waiting.

He didn't sleep much.

Dawn came clear and cool. The men stood to at first light, rifles ready. But nothing happened. The sun rose. The valley filled with pale light. Birds sang in the olive trees.

They stood down after an hour and ate breakfast. Hard biscuits and tea. Dick's hands shook slightly holding his mug. Not fear. Just tension, waiting, knowing something was coming.

"Steady," Tom said.

Dick nodded and drank.

The morning wore on. Nothing happened. Men cleaned weapons, talked quietly, and waited. The sun climbed. The heat built.

At half past seven, Sergeant Harris came through. "Aircraft approaching from the north. Large formation. Stand to."

They were already standing, rifles ready. The sound came first. A low rumble that built and built. Then they appeared. Dozens of aircraft flying in formation, wings catching the sun.

"Junkers," someone said. "Ju 88s."

The aircraft kept coming. More and more of them, filling the sky. Dick had never seen so many planes at once. They circled but didn't dive yet. Just circled like predators deciding where to strike.

"Here they come," Mick said.

The first aircraft peeled off and dived, engine screaming. The whistle of bombs followed, high and thin.

Then the valley erupted.

Explosions walked across the airfield, throwing up fountains of earth. The blast waves hit seconds later, pressure against chest and ears. More planes dived. More bombs fell. The noise was constant, enormous, beyond anything Dick had experienced.

He pressed himself into his trench, hands over his head. The ground shook. Dirt rained down. Through the noise he heard men shouting though he couldn't make out words. The bombing went on and on, each wave dropping their loads before climbing away.

When it stopped, the silence shocked him. His ears rang. Smoke drifted across the valley. The airfield was cratered, several parked aircraft burning.

"Everyone all right?" Sergeant Harris moved along the line. "Sound off."

They called their names. Everyone in Dick's section had survived. Down the slope, stretcher bearers moved toward the aid station.

Then someone shouted. Pointed at the sky.

More aircraft are coming. Different this time. Transport planes, flying low and slow, engines laboring. Twenty or more, approaching from the north in line abreast.

"What are they doing?" Young Harris asked. "They're too low for bombing."

"Wait," Sergeant Harris said. "Just wait."

The rear doors opened. Small shapes tumbled out into the air. Dark bundles falling, spinning, dropping.

Then the parachutes bloomed.

White canopies. Dozens of them. Hundreds. Opening against the blue sky like enormous flowers. The bundles beneath took shape as they descended.

Not supplies. Not equipment.

Men.

"Christ," someone whispered. "They're men."

"They're dropping soldiers," Tom said, his voice strange. "They're dropping soldiers from planes."

Dick stared. Men falling from the sky under silk

canopies, drifting down toward the airfield and the olive groves. It was impossible. Armies didn't do this. Armies couldn't do this.

But they were doing it. The sky filled with parachutes. More transport planes coming, more men tumbling out.

"What do we do?" someone asked.

No one answered. They all watched, unable to look away.

Then Sergeant Harris's voice cut through. "They're in the air! They can't shoot back! Fire! Fire at will!"

Dick raised his rifle, aimed at the nearest parachute, and squeezed the trigger. Around him the slope erupted with gunfire. The Bren guns opened up, their chatter cutting through the crack of rifles.

The Germans hung in their harnesses, exposed, unable to dodge. Dick saw men jerk and go limp. Saw parachutes collapse, lines cut by bullets. Saw bodies falling faster, tumbling without the canopy to slow them.

But there were so many. More transports kept coming. More parachutes kept blooming. The wind caught them, scattered them wide. Some landing in the olive groves, some on the slopes, some on the airfield.

Dick reloaded. His hands moved automatically but his mind was still trying to understand what he was seeing. Men falling from the sky. Like something from a myth.

"Keep firing!" Harris shouted. "Don't let them reach for their weapons!"

Dick aimed at another parachute and fired. The man jerked. Dick fired again and the parachute lines parted.

The German plummeted.

The ones who reached the ground alive cut free and ran for cover. Dick saw they wore grey-green uniforms and carried no weapons. The weapons were in separate containers, dropping on their own parachutes. The Germans ran across open ground, exposed, desperate to reach the containers.

Dick shot a man running toward a container. The man fell. He aimed at another struggling to open a container. The man went down before he could pull out his rifle.

More transports roared overhead. More parachutes filled the sky. Dick's shoulder ached from recoil. Sweat ran into his eyes. The air was thick with cordite. And still the Germans came, falling from the sky like rain, like snow, like nothing anyone had seen before.

"They're mad," Young Harris said, feeding a magazine into the Bren. "They'll lose half their men before they touch the ground."

"They know that," Mick replied, firing in controlled bursts. "They're accepting it. They think they can win anyway."

"Can they?"

No one answered. They kept firing.

A paratrooper landed in the olive grove twenty yards below Dick. The man fought out of his harness, looked around, started running uphill. Dick tracked him, squeezed the trigger. The man went down hard.

Dick ejected the spent cartridge, chambered another round. He'd just shot a man, watched him fall. No time to think about it. Just load, aim, fire. Load, aim, fire.

The morning blurred into continuous combat. Germans who reached the ground and found weapons gathered in groups and began to move. Some headed for the airfield. Others moved up the slopes toward the defensive positions. They were well trained, aggressive, using cover, advancing by bounds.

These weren't ordinary infantry. These were elite troops, specially trained. And they fought with the desperation of men who'd been dropped into hell and had to fight their way out or die.

Dick's section held their position, pouring fire down into the olive groves. The Bren hammered away. Young Harris fed in magazines, calling out targets. Tom was on Dick's left, firing methodically, taking his time.

A group of Germans made it to the trees below them. Dick could see them moving through the dappled shade. Four or five men, working uphill. They were young, faces streaked with dust and sweat. One looked up and Dick saw his expression. Fear and determination mixed together.

Dick fired. The man dropped. The others scattered, returning fire. Bullets cracked overhead, clipping through olive branches. Dick ducked, reloaded, came up and fired at another German moving between trees.

"Grenade!" someone shouted to the left.

The explosion came from where Walsh's fire team was positioned. Then more firing, shouting, the sound of men fighting at close range.

"Hold the line!" Sergeant Harris was somewhere behind them. "Don't let them break through!"

The fighting went on through midday. The sun climbed high and beat down. Dick's canteen ran dry. His ammunition was running low. His shoulder was bruised from recoil, his hands cramped from gripping the weapon.

But they held. The Germans in front of their position were pinned or dead. Transports tried to land on the airfield and were shot to pieces. The runway was littered with crashed aircraft and burning wreckage. Dead paratroopers hung in the olive trees, parachutes draped like shrouds.

Around two o'clock, Sergeant Harris crawled into their position. His face was grey with dust and fatigue.

"Ammunition?"

"Low," Dick said. "Maybe twenty rounds per man. Bren's got one magazine left."

"Same everywhere." Harris looked down at the carnage spread across the valley. "They're bringing up reserves. We've hurt them badly but they keep coming."

"Are we holding?" Tom asked.

"For now." Harris pulled out a crumpled map. "If we have to pull back, you head south toward that ridge." He pointed to hills visible in the distance. "Keep your men together. The Stukas will be back."

After he left, Dick redistributed the ammunition. Everyone got at least ten rounds. Not much. If the Germans came again in strength, they'd run dry in minutes.

"How's Walsh's team?" Mick asked.

"Corporal down," Dick said. "Shrapnel. They pulled him

back."

The afternoon dragged on. The firing died to sporadic shots. The Germans were regrouping, bringing up more men and supplies. Dick could see them moving at the western end where they'd established a foothold. Transports were landing there now, taxiing quickly before taking off again.

By evening, the first day's attack had been repulsed. The Germans held ground at the western edge but had paid for it. Bodies lay everywhere. German and Allied both.

Dick sat in his trench as the sun set, too tired to move. His rifle lay across his lap. His hands were black with powder residue. He'd lost count of how many shots he'd fired, how many men he'd hit. The day had passed in continuous present. Moment after moment of loading and firing and reloading. No space for reflection.

Tom crawled over and sat beside him, offering his canteen. Dick drank, the water warm and tasting of canvas but precious.

"You all right?" Tom asked.

"I don't know."

Tom was quiet. Then he said, "Walsh is dead. Died on the stretcher before they got him to the aid station."

Dick nodded slowly. Walsh. The quiet corporal. Dead now. Just dead. The space where he'd been was empty.

"There'll be more tomorrow," Tom said. "You know that."

"Yes."

They sat together in silence as darkness fell. Other men

did the same. Sitting, drinking what water they had, checking weapons, trying not to think about the next day.

But the next day came anyway.

The Germans attacked at dawn. More transport planes, more paratroopers falling from the sky. The fighting was just as intense, just as close. Dick fired until his rifle barrel burned. His ammunition ran low. They drew more from dwindling reserves.

The Germans brought in gliders. Large silent craft that landed on beaches and in olive groves, disgorging men and equipment. They brought in mountain troops, elite units that moved through broken ground with practiced ease. More reinforcements, landing on the captured portions of the airfield despite the fire.

The Allies were running out of everything. Ammunition, water, food, men. The defensive line stretched thin. Units got isolated. Communications broke down. Orders were confused or contradictory.

On the third day, Dick's section was pulled back and redeployed closer to the airfield. The Germans had broken through on the western side, threatening to encircle the New Zealand positions. The fighting was closer now. Sometimes hand to hand in the olive groves, sometimes shooting at men only yards away.

Dick saw Young Harris go down, hit in the chest by rifle fire. The medics dragged him away but Dick saw his face. Grey. Already dying. Young Harris who'd cleaned his rifle three times that first day.

Mick took over the Bren alone, firing it from the hip as they fell back through the trees.

By afternoon of the third day, the battle was lost. The Germans owned the western half of the airfield, bringing in reinforcements faster than the Allies could stop them. Artillery landing now. Half-tracks. Supplies that would let them push inland.

Dick gathered his section. Tom and Mick. Two others from the replacements. Five men out of eight who'd started.

"We're falling back," Dick said. "South toward the ridge. Stay spread out. Watch for aircraft. If we get separated, head for high ground and link up after dark."

They moved through the olive groves, keeping low, using cover. German patrols everywhere. Twice they went to ground while enemy soldiers passed within twenty yards. The second time, a young German private stopped and looked directly at where Dick lay behind a stone wall. Their eyes met.

The German turned away and kept walking.

Dick waited until they were gone, his heart hammering against his ribs. He didn't know if the German had seen him and chosen not to raise an alarm, or if the shadows had hidden him. Either way, he was grateful.

They reached the ridge as the sun set. From there Dick looked back toward Maleme. Smoke hung over the valley. Aircraft landing continuously now, a steady stream. The battle was over. The airfield was lost. Crete would fall.

That night they dug in on the ridge and waited. Other soldiers appeared from the darkness. Individuals, small groups, all heading south. Some were wounded. Most were just exhausted. They gathered in loose clusters, sharing what little they had.

Around midnight, an officer appeared. A captain Dick didn't recognize, his uniform torn, his face haggard.

"New orders. We're withdrawing to the south coast. Ships at Sphakia for evacuation. About thirty miles. We move tonight, rest during the day, move again tomorrow night. Stay in small groups. Avoid roads. The Germans own the air."

"What about the wounded?" someone asked.

The captain hesitated. "Do what you can. But the mission is to get as many men off this island as possible."

After he left, Dick looked at his section. They were all watching him, waiting.

"We move in an hour," Dick said. "Get what rest you can. Check your kit. We travel light. Weapons, ammunition, water. Everything else stays."

They dozed in shifts. Real sleep was impossible. Dick sat with his back against a tree and thought about the past three days. About Walsh dead with shrapnel in his chest. About Young Harris shot down in the olive grove. About the Germans falling from the sky and the noise and the blood and the simple mechanics of killing.

He thought about being responsible for these men. He'd never asked for it. It had simply happened, through attrition and accident. But now he had it. The weight of keeping them alive long enough to reach the coast.

When the hour was up, they moved out. Five men in the darkness, navigating by stars, heading south toward a coast they couldn't see and ships that might not be there.

Egypt, Greece, now Crete. Three battles. Three defeats. Three retreats.

The pattern was clear.

Dick didn't know how much longer he could keep doing this. But he would do it anyway, because that's what soldiers did. They kept going until they couldn't.

The withdrawal had begun.

CHAPTER 7 — WITHDRAWAL

They moved through the night in single file, twenty yards between each man. Dick led, using the stars to navigate south. Behind him came Tom, then Mick with the Bren gun, then the three replacements. Walsh, Cooper, Jenkins. Men from different units, thrown together by circumstance.

The track wound through scrub and low hills, climbing and descending in patterns that made Dick check the stars repeatedly to confirm they hadn't turned in circles. Animal path, not road. Loose stones shifted underfoot. Thorny bushes caught at uniforms.

Dick's sweat-damp uniform chilled him as they walked. His boots, worn through at the heels since Egypt, let him feel every stone. Blisters forming with each step. Nothing to do but keep walking.

They stopped each hour to rest and listen. Around two in the morning, engines in the distance. Trucks on one of the main roads, moving with headlights. German trucks. The roads belonged to them now.

"How far?" Tom asked quietly.

"Maybe five miles."

"Twenty five to go then."

Dick said nothing. More like thirty five, maybe forty if they had to avoid patrols. Tom knew that.

"We're not going to make it." Tom's voice was flat, matter of fact.

"We'll make it."

"Dick."

"Then we die trying." Dick stood. "Come on."

They walked.

Dawn came with the sky pale in the east. They found shelter in a dry creek bed overhung with scrubby trees. Not perfect cover, but enough if they stayed still.

Dick shared the water. His canteen was three quarters empty. The others are no better.

"Mountains like these always have springs," Walsh said. He was the oldest of the replacements, maybe thirty, with weathered hands. "We just have to find one."

"We'll look as we go," Dick said. "But we can't waste time searching."

They slept in shifts, two men on watch. Dick took first with Mick. They sat at opposite ends of the creek bed, rifles ready.

The sun climbed. Even in the shade the heat built. Dick allowed himself one mouthful of water and forced himself to wait before taking another. His throat stayed dry, his lips cracking.

At midday, Stukas passed overhead. Dick pressed himself flat and didn't move. The planes circled once, then flew west. After they were gone, Dick found he'd been holding his breath.

Late afternoon they moved again. The track climbed steadily. In the distance, mountains rose with patches of snow still clinging to their peaks. They'd have to cross those mountains or go around. Either way would take time they didn't have.

As dusk fell, gunfire. Sharp bursts, not close but not far enough. Someone fighting or being hunted.

"Keep moving," Dick said. "Don't stop."

They pressed on, angling away from the sound. The light faded. Stars appeared. Dick found the Southern Cross and checked their direction.

The ground climbed more steeply. The track narrowed between boulders and stunted trees. Behind Dick, Mick's breathing grew harsh. The Bren gun taking its toll. Dick thought about telling him to leave it but knew Mick wouldn't.

They climbed for an hour, maybe more. The path switched back up a slope covered in loose scree. Every step sent stones rattling down. The sound carried.

Halfway up, Dick stopped. They gathered beside a boulder, all gasping.

"Water," Cooper said. "I'm out."

"Me too," Jenkins added.

Dick checked his canteen. Two mouthfuls left. Tom's the same. Walsh shared what he had, a few drops for each man.

"We'll find water tomorrow," Dick said.

They rested for ten minutes, then moved on. Dick's calves burned. His breath came ragged. Behind him the others struggled but no one complained.

The slope began to level. They were nearing the top.

The first shot cracked out.

Dick dropped flat, the rifle's echo slamming off rocks. Behind him, the others hit the ground. Silence. Then another shot, stone chips spraying near Dick's head.

"Get to cover!"

They scrambled backwards, sliding on loose scree, grabbing at rocks. More shots came, muzzle flashes visible higher up the slope. Germans. A patrol or checkpoint.

Dick found shelter behind a boulder and brought his rifle around. He couldn't see the Germans clearly, just shapes, but he aimed at the muzzle flashes and fired. The rifle kicked. He worked the bolt, fired again.

Tom was firing beside him. Mick had the Bren gun set up behind a rock outcrop, laying down short bursts.

The Germans returned fire. Bullets snapped overhead, ricocheted, sent rock splinters flying.

"We need to pull back!" Tom shouted.

"Do it! One at a time! I'll cover!"

Tom scrambled backwards, half sliding. Then Jenkins. Then Cooper. Dick and Mick kept firing, keeping the Germans" heads down.

"Go!" Dick shouted to Mick.

Mick gathered up the Bren, lifting it with a grunt. He started down the slope, bent over, the gun clutched to his chest.

Dick fired his last rounds, then turned and followed. He slid more than climbed, his boots losing purchase. Bullets cut through the air around him.

Then a different sound. A single sharp crack.

Mick cried out.

Dick looked back. Mick stood upright twenty feet above him. The Bren gun slipped from his hands. He swayed once, then started to fall.

Dick scrambled back up. Mick was on his side on the stones, one hand pressed to his ribs. Blood welled between his fingers, black in the starlight.

"Mick!"

Mick looked at him. His eyes were wide, confused. He tried to speak but only air came out, a wet rattling. His hand lifted halfway, reaching for something that wasn't there, then fell. His weight shifted and Dick caught him, easing him down.

The blood spread across Mick's shirt. Dick pressed his hand over the wound but there was too much. It pulsed between his fingers with each heartbeat, hot and slippery. Mick's face was already pale, his lips losing color.

"Stay with me," Dick said. "Mick. Stay with me."

Mick's eyes focused on him for a moment. His mouth moved, forming words Dick couldn't hear. Then his eyes went empty. His body went heavy in Dick's arms. His chest didn't rise again.

"Mick!" Dick shook him. Nothing.

Tom was there, grabbing Dick's arm. "He's gone! We

have to go!"

"I can't leave him!"

"He's dead, Dick! We have to move!"

More shots from above. The Germans were advancing, working down the slope.

Dick looked at Mick one more time. At his face, still and peaceful. At the Bren gun lying beside him. Then Dick grabbed the gun, slung it over his shoulder, and scrambled down after Tom.

They caught up with the others at the base, pressed into shadow. Dick was breathing so hard he thought his lungs would burst. Mick's blood was on his shirt, warm and sticky.

"Where's Mick?" Walsh asked.

Dick shook his head.

"Christ," Tom said quietly.

They waited in shadow, listening. The firing above had stopped. Either the Germans thought they'd killed everyone, or they were being cautious about pursuing in darkness.

"We go east," Dick said, his voice rough. "Circle around. Then head south when we're clear."

They moved east, staying low, using cover. The Bren gun dug into Dick's shoulder. He understood now why Mick had struggled. But he wouldn't leave it. Mick had died carrying it.

They walked through the night, wide of the checkpoint. Once they heard voices. German voices, calling in the darkness. They froze and waited until the voices faded.

By dawn they were miles from where Mick had died. They found shelter in another dry creek bed, deeper, better concealed. Dick set the Bren down and sat with his back against the bank. His shoulder was bruised and raw where the gun had rested.

The others collapsed around him. Dick closed his eyes and saw Mick's face. The surprise in his eyes. The blood.

Jack and Mick. Both gone.

"We need water," Walsh said. "By tonight or we're finished."

"I know."

"There might be a village. Could have a well."

"Could have Germans."

"We don't have a choice."

Walsh was right. They were out of water, out of food, nearly out of time.

"We'll scout it this afternoon," Dick said. "If it looks clear, we'll get water. If not, we keep going."

In the afternoon, Dick and Walsh reconnoitered. They found a village half a mile east. White stone houses around a central square. A well. Washing on lines. Chickens in the dirt.

But German soldiers too. Half a dozen, sitting in shade, rifles stacked. Relaxed.

"Can't risk it," Walsh whispered.

They crept back and told the others. No water.

As the sun lowered, they prepared to move. Dick checked the Bren gun, making sure it was ready. Load the magazine, pull the cocking handle, squeeze the trigger. Simple enough.

They moved out at dusk, heading south. The land rose steadily into real mountains now. The air cooler, stars brighter.

Dick's legs threatened to buckle with each step. His tongue was swollen. His vision had narrowed, dark at the edges. Behind him the others were no better. Jenkins stumbling every few steps. Cooper's breathing labored. Even Walsh was moving like an old man.

Around midnight they heard running water.

At first Dick thought he was hallucinating. But Tom heard it too, and Walsh, and they all turned toward it.

A spring, bubbling up between rocks, running through a narrow channel before disappearing underground. Clear and cold, tasting of stone. Cold enough to hurt his teeth.

They drank until their stomachs hurt. Filled their canteens. Splashed water on their faces.

"We'll rest here," Dick said. "An hour."

They sat by the spring. Dick's mind cleared. His vision sharpened. He could see individual leaves on the scrubby trees now. His body still ached but at least it had fuel.

After an hour they moved on. The track continued climbing. Mountains rose around them, dark shapes against stars. Somewhere beyond those mountains was the sea.

They climbed through the night. The path grew steeper, rockier. Several times they had to haul each other up sections that were more cliff than path.

The Bren gun ground against Dick's shoulder. He shifted it to the other side, then back when that one started to hurt.

Around three in the morning, aircraft passed overhead. Multiple engines, heading south. Bombers, probably. Heading for the evacuation beaches.

"They're bombing the coast," Tom said quietly.

Dick said nothing. They kept moving.

Dawn found them high in the mountains, above the treeline. Bare rock and thin soil. They took shelter in a crevice between boulders and looked back.

Below them lay northern Crete. Dick could see the coast, the sea beyond. Smoke rising from multiple points. Maleme. Suda Bay. Other places.

"We must have come fifteen miles," Walsh said.

"Still got another ten or more," Tom replied.

At midday Walsh shook Dick awake. "Movement. On the track we came up."

Dick crawled to the edge and looked. Far below, maybe a mile, figures moving. A column, working up the mountain path. Moving slowly, methodically, stopping to search the ground.

"Germans," Dick said. "Tracking us."

"How close?"

"Two hours. Maybe less if they move fast."

Dick looked at the others. Exhausted, running on water and will. The Germans were fresh, supplied, armed.

"We move now," Dick said. "Put distance between us."

They gathered their kit and set off. The track continued upward through high passes where wind cut like a knife. Dick's lungs burned. His legs shook.

But he kept moving. They all did.

The sun climbed. Around mid afternoon they crested a high pass and saw the southern coast. The sea. Blue, stretching to the horizon. And on the beaches, tiny shapes that might be ships.

"There," Tom said, pointing. "That must be Sphakia."

Five miles as the crow flies. But they weren't crows. Eight miles, maybe ten, descending the mountain, working through foothills, circling German positions.

"Let's go," Dick said.

They started down. The descent was harder, the loose scree treacherous. Jenkins fell twice, tearing his trousers. Cooper twisted his ankle and had to limp.

As they descended, more aircraft. Stukas, diving on the beaches. The whistle of bombs. Columns of smoke rising.

By evening they were in the foothills, the coast visible through gaps in terrain. So close now. Dick could almost taste the salt air.

That was when they ran into the roadblock.

Two German soldiers beside an overturned cart across the road. They saw Dick's group before Dick saw them.

"Halt!" one shouted in German, then broken English. "Stop! No move!"

Dick froze. The two Germans stood, raising rifles. One

shouted back over his shoulder. More soldiers appeared from a building. Four of them. Six total.

Too many. No ammunition. No cover.

Dick looked at the others. At Tom, who'd been with him since Egypt. At Walsh and Cooper and Jenkins, who he barely knew. At the Bren gun on his shoulder.

He thought about running. About dying with a rifle in his hand. But that would get them all killed.

Dick lowered the Bren gun to the ground. Then he raised his hands.

"Don't shoot," he said. "We're done. We surrender."

The others lowered their weapons and raised their hands. The Germans approached, rifles trained. One, an older man with corporal's stripes, gestured for them to kneel.

They knelt on the dusty road, hands on heads. The Germans collected their weapons, searched them. Dick had nothing but his identification tags and the photograph of his family. The German who searched him looked at the photograph, nodded, and handed it back.

"You keep," the man said. "Family is important."

Dick took the photograph and put it in his pocket. His hands were shaking.

They sat by the roadside while the Germans waited for transport. The two checkpoint soldiers shared water with them. Not much, but some. They looked nearly as tired as Dick's group, faces drawn, uniforms dusty.

"War shit, yes?" one said in broken English.

Dick nodded. "Yes. War shit."

The German smiled sadly and went back to his post.

As the sun set, a truck arrived. They were loaded into the back with other prisoners. Australians, New Zealanders, a few British. No one spoke much.

The truck drove north, back toward Maleme, back toward the airfield they'd failed to hold. Dick sat with his back against the truck's side and looked at the mountains receding behind them. The mountains they'd crossed. The withdrawal they'd survived.

He thought about Mick and Jack. About Bluey, wounded and lost. About all the others.

He reached into his pocket and pulled out the photograph. His family looked back at him from another world. His mother and father. His sisters. Lorna, who would read his letter and understand.

Dick looked at the photograph for a long time. Then he folded it and put it back in his pocket, next to his heart.

Whatever came next, he would keep it.

The truck rumbled on through the darkness, carrying him away from the sea, away from freedom. His war as a soldier was over.

His war as a prisoner was beginning.

CHAPTER 8 — CAPTURE

The German corporal who'd handed Dick back his photograph gestured for them to sit by the roadside. They sat, hands on their heads, while the soldiers discussed what to do with them. German, incomprehensible, but the tone was matter of fact. Not hostile. Not friendly. Just soldiers doing a job.

Dick's legs trembled. Now that the walking had stopped, everything was catching up. His vision doubled, merged, doubled again. He looked at Tom beside him, at Walsh and Cooper and Jenkins beyond, and saw them swaying where they sat.

Five miles. Maybe less. He could still see the mountains they'd descended. The sea was beyond the next ridge. Ships had been there.

"Water?" one of the German soldiers asked in English, offering his canteen.

Dick nodded. The German passed the canteen. Dick drank a small amount and passed it to Tom. The water tasted of iron but it was water.

The German who'd given it couldn't have been more than nineteen. Blond hair, blue eyes, face burned red by the sun. He looked at Dick.

"You come far?" he asked, his English halting.

"Maleme," Dick said.

The German nodded. "Long walk. Hard walk. We also walk a lot. Very tired."

No triumph in his voice. Just one tired soldier talking to another. Dick wondered if this boy had jumped from one of those transport planes. If he'd been one of the parachutes Dick had shot at.

A truck arrived twenty minutes later. An old flatbed with wooden sides. The German corporal ordered Dick's group into the back. They climbed in stiffly and found a dozen other prisoners already there. Australians mostly, a few New Zealanders, two British soldiers.

No one spoke. They sat on the floor and stared at nothing. Dick recognized one of the Australians, a private from another company he'd seen at Maleme. The man's face was grey with dust, his eyes hollow.

The truck drove north. Back the way Dick had come. All that walking south. Now driving north.

The sun climbed. The wooden sides radiated warmth like an oven. Dick closed his eyes.

The truck stopped twice. Once to let a German convoy pass. Once at a checkpoint where guards counted prisoners and waved them on. Each time Dick looked at the landscape passing by and thought about running. But his body wouldn't carry him ten yards.

They reached Maleme late afternoon. The airfield had transformed. Transport planes landing and taking off in a steady stream. Tents and temporary buildings. Trucks and half-tracks on newly graded roads. The

Germans had turned the battlefield into a base in days.

The truck pulled up beside a large fenced compound near the eastern edge. Inside the fence, hundreds of prisoners sat or lay on bare ground. Allied soldiers, collected from across the island. The walking wounded. The exhausted. The defeated.

"Out," a German guard ordered in English. "Move."

Dick climbed down, his legs nearly giving way. Tom steadied him. They walked together toward the compound gate. German soldiers with rifles stood at intervals along the fence, watching. They looked tired too, faces drawn, uniforms dusty.

Inside, Dick and the others were directed to sit. No shelter, no shade. Just open ground with the sun beating down. Dick sat and looked around.

He recognized faces. Men from his battalion, from other battalions, from units he'd seen during the retreat. All wore the same expression.

A New Zealand sergeant sat nearby, his arm in a crude sling. Dick caught his eye.

"How long have you been here?" Dick asked.

"Two days," the sergeant replied. His accent was broad, rural. "Caught on the third day. Been sitting here since."

"Any word on what happens next?"

"Germany, probably. Prison camps. Could be worse."

Dick nodded.

Tom sat down beside him with a groan. "Well. This is it then."

"For now."

"You think there's a later?"

"There's always a later," Dick said. "Until there isn't."

Tom managed a tired smile. "Philosopher now, are you?"

"Just too tired to lie."

They sat in silence as the sun moved across the sky. More trucks arrived with more prisoners. The compound filled. By evening there must have been four or five hundred men in a space meant for half that.

German soldiers brought water in large containers. Prisoners lined up with canteens, cups, helmets, and were given a ration. Not much. Then bread, hard and stale, and tins of something that might have been meat. Dick ate without tasting, just filling his stomach.

As darkness fell, the temperature dropped. Dick had no blanket, no coat. He sat with his arms wrapped around his knees and shivered. Around him, other men did the same. Some talked quietly. Most sat in silence.

Dick pulled out the photograph of his family. In the fading light he could barely make out their faces. His mother's gentle smile. His father's stern expression. His sisters, all five of them.

Lorna would be eighteen now. What would she say if she could see him? Sitting in the dirt behind barbed wire. He thought about the letter he'd written to her. She would understand. She always did.

"Is that your family?" Walsh asked, sitting down beside him.

"Yes."

"They're waiting for you. Remember that."

Dick nodded and put the photograph away. His family was ten thousand miles from here, living their lives. They probably didn't know yet. By the time word reached them, he could be anywhere. Germany, Poland, somewhere he couldn't imagine.

The night deepened. Guards moved along the perimeter, dark shapes against darker sky. Somewhere in the distance, an aircraft engine coughed to life. The war continued.

Dick lay down on the hard ground and closed his eyes. His body was exhausted but his mind kept replaying the past week. The battle at Maleme. Jack dying. The withdrawal. Mick falling on the slope, blood spreading across his shirt. The desperate trek. The roadblock.

So many small moments where things could have gone differently. If they'd taken a different path. If they'd moved an hour earlier. If the German patrol hadn't been at that exact spot.

Eventually exhaustion won and he drifted into half-sleep where consciousness blurred and time passed without meaning.

He woke at dawn to shouting. German orders, harsh and commanding. The prisoners stirred, getting to their feet, forming loose groups. Dick stood, every muscle protesting. His mouth was dry, his stomach empty despite the bread.

Guards moved through the compound, counting prisoners, organizing them. A German officer appeared, a captain with a lean face and sharp eyes. He stood on a box so everyone could see him and spoke in heavily accented English.

"You are prisoners of war. You will be treated according to the Geneva Convention. You will be given food and water. You will be given medical care if needed. You will follow orders. Anyone who tries to escape will be shot. Anyone who causes trouble will be shot. This is war. We did not make these rules, but we will enforce them. Is this understood?"

A murmur of acknowledgment.

"You will be moved from here in groups. Some today, some tomorrow, some the day after. You will go to transit camps, then to permanent camps in Germany. The war for you is over. How long you remain prisoners depends on how long the war lasts. That is not my decision or yours. It simply is what it is."

He stepped down and walked away.

"Geneva Convention," Tom muttered. "That's something, at least."

"If they follow it," someone else said.

"They will," the New Zealand sergeant said. "Germans are bastards, but they're not animals. They'll follow the rules. Mostly."

The day dragged with the heat. Prisoners sat in whatever shade they could find. More water was brought, more bread. Some men were called forward for medical inspection. Those with serious wounds were taken away.

Dick sat and stared at the dirt between his feet. Tom asked him something at one point but Dick didn't answer. He'd heard the words but they hadn't meant anything.

"You all right?" Tom asked again.

Dick looked at him. "I don't know."

"We're all in shock. It takes time."

"How much time?"

Tom shrugged.

In the afternoon, the first group was called forward for transport. Fifty men, alphabetically by surname. Dick's name wasn't called. Neither was Tom's. They watched the men loaded onto trucks and driven away.

The sun set. Dick lay on his back and looked at the stars. The same stars from the trenches. The same stars that had guided him during the withdrawal. They were there, same as always.

On the second day, more names were called. This time Dick heard his own. He stood, gathered his canteen, his identification tags, the photograph, and moved toward the gate with Tom and Walsh. Cooper and Jenkins had been called in an earlier group.

They were loaded onto trucks, forty men per vehicle. The trucks pulled out and headed north along the coastal road. Dick watched the airfield recede, watched the mountains in the distance, watched Crete disappear.

The truck drove through the afternoon and into the evening, stopping twice for prisoners to relieve themselves by the roadside. German guards watched carefully but not cruelly.

As night fell, they reached a larger compound. A proper camp with barracks and fencing and guard towers. A transit facility, the German officer explained, where they'd be processed before being sent to permanent camps in Germany or Poland.

They were herded into a large building and told to strip. German soldiers searched their clothes while others examined the prisoners for injuries or disease. Humiliating but efficient. Within an hour they had their clothes back and were assigned to barracks.

The barracks were crowded but had roofs and wooden bunks. Dick claimed a spot on a lower bunk. For the first time in nearly a week he'd had anything resembling a bed.

Tom took the bunk above him. "Well," Tom said, looking around, "home sweet home."

"For now," Dick replied.

"For now," Tom agreed.

They were given soup that night, thin and tasteless but hot. And more bread, slightly fresher. Dick ate slowly, making it last.

After the meal, prisoners were allowed outside for an hour before lights out. Dick stood in the yard and looked up at the sky. The stars were there, same as always. He thought about Newcastle. About his family sitting down to dinner. About Lorna arguing with Jean about something trivial. About his mother's quiet voice bringing peace to the table.

He wondered if they knew yet. If the telegram had arrived. *We regret to inform you that your son, Corporal James Owen Roberts, is missing in action and believed captured.* Or maybe just missing, because the Germans might not have reported the prisoners yet. His family might still think he was fighting. Or dead. Or lost somewhere.

He hoped they knew he was alive. At least he was alive.

A whistle blew. Time to return to barracks. Dick walked back inside with the others and climbed onto his bunk. The lights went out. The room filled with the sounds of men settling, coughs, whispers, the creak of wood.

Dick lay in the darkness and closed his eyes. Tomorrow they'd probably be moved again. Put on trains or trucks and sent north to Germany. The war would continue without him. The world would keep turning.

He slept without dreaming, his body finally getting the rest it needed. And when he woke, it would be to a new life. A harder life. A life measured not in battles won or lost but in days endured.

His war as a soldier was over. His war as a prisoner had begun.

CHAPTER 9 — TRANSPORT NORTH

The whistle blew before dawn, sharp and insistent. Dick woke to shouting, German voices echoing through the barracks, boots on wooden floors. Around him, men stirred and groaned, their bodies stiff from thin mattresses.

"Up! Raus! Schnell!"

Guards moved down the aisle between bunks, banging rifle butts against wooden frames. The electric lights flickered on, harsh and yellow. Dick sat up, his head nearly hitting the bunk above. Tom was already moving, swinging his legs down.

"Here we go then," Tom said, his voice thick with sleep.

Dick rubbed his face. He'd been dreaming about home. Newcastle. The beach. It slipped away.

Outside, the air was cold enough to see their breath. The sky was still dark, just a thin line of grey to the east. Hundreds of prisoners stood in ragged lines, shoulders hunched, faces blank.

A German officer stood on a wooden crate and announced in accented English that groups would be called for transport. Names read alphabetically. When called, prisoners would collect their belongings and report to trucks outside the main gate.

Dick had no belongings except what he wore and carried in his pockets. The photograph of his family, his identification tags, a stub of pencil he'd found during the march. That was everything he owned now.

The officer began reading names. British soldiers first. Then Australian names started appearing. Dick heard his own called third.

"Roberts, James Owen, Corporal."

"Here," Dick called back.

"Wilson, Thomas Edward, Private."

"Here," Tom said beside him.

Walsh's name came moments later. The three of them moved forward together. Dick looked back at the barracks they were leaving, at the other prisoners still waiting, and felt relief and dread in equal measure. At least something was happening. But moving meant getting further from home, deeper into Germany.

They waited by the gate for nearly an hour while more names were called. A few New Zealanders, more Australians, some British. By the time the count was complete, there were perhaps two hundred men standing in the cold dawn, stamping their feet.

German guards distributed the men among three trucks. Dick, Tom, and Walsh were assigned to the second. They climbed into the back, joining thirty other prisoners in the covered cargo bed. No seats, just metal floor and canvas sides that let in drafts of cold air.

As the truck started moving, Dick found a spot near the back where he could lean against the frame. Tom sat beside him. Walsh wedged himself in nearby.

Through the gap in the canvas, Dick could see countryside passing. Fields, mostly empty now in late May. A few farmhouses with smoke rising from chimneys. Trees in full leaf. It looked peaceful, ordinary.

They drove for two hours before stopping. The guards let them out to relieve themselves by the roadside and gave them water from jerry cans. The water tasted of metal and chemicals, but Dick drank it anyway.

A German guard, young and fair haired, stood nearby watching. He didn't look hostile, just bored. Dick caught his eye and the guard nodded slightly.

After fifteen minutes they were ordered back into the trucks.

The day wore on. The sun climbed higher and the temperature inside the truck rose. Men dozed, leaning against each other. Some talked in low voices. Most stayed quiet.

Dick's mind drifted to Newcastle. To the beach and the sound of surf. To his mother's kitchen and the smell of dinner cooking. To his sisters laughing while they set the table.

He thought about Eileen standing under the pharmacy awning. Her offer to get engaged. His refusal. Maybe she'd been right. Maybe he should have said yes.

"Your family?" Walsh asked quietly, nodding at the photograph Dick had pulled from his pocket.

"Five sisters. Parents."

"That's a lot of women to manage."

"My mother manages them. I just tried to stay out of the way."

Walsh smiled slightly. "I've got three brothers. The house was always loud. Boys fighting about everything. I used to wish for some quiet." He gestured at the truck, at the silent men. "Be careful what you wish for."

Dick put the photograph away.

Around midday they stopped at a railway siding. A freight train waited on the tracks, long and grey, its engine idling. The prisoners were ordered out and formed into lines beside the wagons.

Dick looked at the train. Freight wagons. Cattle cars.

German guards moved along the line, counting men and directing them toward different wagons. Forty men per wagon, someone said. Maybe fifty if they packed them tight.

Dick's group was directed to the fourth wagon from the front. He climbed through the open door into the dim interior. The wagon was empty except for thin straw on the floor and a single bucket in the corner. The smell was of old wood and oil and piss.

Tom climbed in after him, then Walsh. More men followed until the wagon was crowded, men standing shoulder to shoulder, no room to sit. When the count reached forty, a guard pulled the heavy door closed. Not all the way, just enough to leave a gap of a few inches.

The darkness was immediate. Dick could barely see the men around him, just shapes and shadows. Someone near the back was breathing hard, on the edge of panic.

"All right?" Tom asked, his hand finding Dick's shoulder.

"Yeah."

"Bloody hell of a way to travel."

"Could be worse."

"How?"

Dick didn't answer.

The train lurched forward, couplings clanging. Then they were moving, wheels clicking over rail joints, the wagon swaying. Men shifted, trying to find positions less uncomfortable. But there was no comfort. The best they could do was lean against each other.

The heat built inside the wagon until the air was thick and hard to breathe. Men stripped off their shirts. The bucket in the corner was used, adding to the smell. Dick counted the clicks of the wheels. He tried to remember the words to songs. He thought about diving into cool water at Newcastle, swimming until his muscles ached.

When the train finally stopped, it was nearly dark. The door was pulled open and fresh air rushed in. Dick gasped. Guards shouted for everyone to get out. Men stumbled and fell as they climbed down, their legs weak from standing so long.

They were given water and allowed to relieve themselves in a field beside the tracks. Dick stood and looked up at the sky. Stars coming out, bright and clear.

They were given food for the first time that day. Hard bread and something that might have been cheese, though it tasted chemical. Dick ate it anyway.

After half an hour they were ordered back into the wagons. The door closed again. The darkness returned. The train moved on.

That night was worse. The temperature dropped and men shivered. The swaying made some men sick. The bucket filled and spilled. Dick tried to sleep standing, his head resting against Tom's shoulder, but real sleep never came.

Dawn came slowly, grey light filtering through the gap. Dick could see faces now, drawn and pale. Tom's eyes were red rimmed. Walsh sat in the corner with his head in his hands.

The train kept moving north. Through the gap Dick watched the landscape change. Flat farmland gave way to low hills. Villages appeared and disappeared. Once they passed a factory with tall smokestacks belching dark smoke.

Around midmorning on the second day, the train stopped at a larger station. When the door opened, German soldiers were waiting with rifles. The prisoners were ordered out and formed into lines on the platform. Dick's legs nearly gave way. He caught himself against the wagon side and waited for his balance to return.

"Count off!" a German officer shouted.

They counted by twos. Forty men from their wagon, all accounted for. The officer consulted a clipboard and divided them into two groups. Dick, Tom, and Walsh were in the second group. They were marched away from the train toward covered trucks parked at the far end of the platform.

They were loaded into the trucks, twenty men per vehicle. There was more room this time, enough to sit. Dick found a spot near the back and sat with his legs stretched out. Tom sat beside him. Walsh was across from them, already half asleep.

The trucks pulled out of the station and onto a main road. Through the back Dick could see the town receding. It looked undamaged, almost normal. People shopping, walking, going about their lives.

They drove for an hour through countryside that was green and rolling. Farmland, mostly, with occasional villages tucked into valleys.

Then the landscape began to change. They passed through an area where fields were torn up, great gouges in the earth. A village appeared that was nothing but rubble, walls blown down, roofs caved in. Dick saw a church with its steeple snapped off halfway up.

The truck slowed. The convoy ahead had stopped. They sat there for twenty minutes while guards walked up and down checking something.

Then they were moving again, but slower now, the trucks in low gear, picking their way carefully. The road was rougher, potholed and broken. Buildings appeared on both sides. Or what was left of buildings. Some were just facades, fronts still standing while backs had collapsed. Others were completely flattened, nothing but bricks and broken timber.

"Where are we?" Tom asked quietly.

No one answered.

The destruction grew worse. Entire blocks were leveled, nothing standing higher than a man's waist. The streets were partially cleared, rubble pushed to the sides to make narrow passages. German soldiers stood at intersections directing traffic. Civilians picked through the ruins.

And then Dick saw them. People. Hundreds of them, maybe thousands. They moved through the ruins like ghosts, clothes torn and dirty, faces gaunt. Men with carts pulling loads of bricks. Women carrying bundles on their backs. Children, thin as sticks, standing in doorways watching the trucks pass.

The truck slowed to walking speed. The ruins pressed close and Dick could see everything. A woman standing in what had been a shop, the counter still there but the walls gone, selling something from a wooden box. An old man sitting on a pile of rubble, his head in his hands. Two boys, maybe eight or nine, digging through a collapsed building with their bare hands.

Dick couldn't look away.

Tom was staring too, his face pale. "Christ," he whispered. "What happened here?"

"Bombs," Walsh said. "Must have been bombed to hell."

The truck turned a corner and Dick saw a square. Trees lined the edges. Or the stumps of trees. In the center was a fountain, dry, its basin cracked and filled with rubble. And around the edges were hundreds of people standing in lines, waiting.

German soldiers were handing something out from the back of a truck. Food, maybe. Or water. The line stretched around the square and down a side street, all of them waiting patiently, no one pushing, all with the same gaunt expressions.

A woman stood at the edge of the square, not in line, just standing. She held a bundle close against her chest, wrapped in cloth. Her coat was torn at the shoulder, one sleeve missing. Her hair was grey and hung loose around her face.

She was looking directly at the truck. Directly at Dick through the gap in the canvas. Her gaze was steady and intense.

Their eyes met. Dick couldn't look away. The woman didn't blink, didn't shift her gaze.

The bundle in her arms moved. Just a small shift. A child. A baby, wrapped against the cold. The woman adjusted her grip, still holding Dick's gaze.

Dick's chest tightened. His throat closed. He couldn't catch his breath properly.

The truck moved on. The woman disappeared from view, but Dick could still see her face.

He turned away from the gap in the canvas. His eyes burned.

Tom put a hand on his shoulder but didn't say anything.

The truck continued through the ruins. Dick made himself look. Street after street of broken buildings. Families living in cellars, their belongings stacked in rubble. Children playing in the wreckage, climbing over broken walls. An old woman sweeping the front step of a building that no longer had a front, just the step and the doorframe and nothing behind it.

He thought about Newcastle. About Alfred Street and the weatherboard houses and the sound of surf at night. About his mother's kitchen, his sisters setting the table, his father coming home from the print shop. About the beach and the parks and the trams running on time.

All of that could be gone. Just like this.

He pulled out the photograph. His family looked back at him from another world. His mother with her gentle smile. His father's stern expression. Jean and Lorna side by side. Dorrie, Shirley, Meryl.

He pressed his thumb against the image.

"That's your family?" a soldier Dick didn't know asked. British private with a Yorkshire accent.

"Yeah."

"They look nice. Safe at home, are they?"

"I hope so."

"They will be. This," he gestured at the ruins outside, "this won't come to England. Won't come to Australia. Too far."

Dick wanted to believe that. But he'd seen the maps. Distance didn't matter anymore.

He folded the photograph and put it back in his pocket.

The convoy left the city and returned to the open countryside. The road improved. The trucks picked up speed.

Dick sat with his back against the truck's side and closed his eyes. But every time he did, he saw the woman with the bundle. Those eyes.

"You all right?" Tom asked quietly.

"No," Dick said. "You?"

"No."

They sat in silence after that. Around them, other men were equally quiet. The whole truck had fallen silent.

Walsh spoke up from his corner. "Anyone know what city that was?"

"Warsaw," someone said. "Had to be Warsaw."

Warsaw. The name settled over them like a weight.

The trucks drove on through the afternoon. They stopped once for water. The guards seemed subdued too.

When they started moving again, Dick dozed. Not proper sleep, just a grey drift. He dreamed about his mother's kitchen. He dreamed about Eileen's face. He dreamed about the woman in Warsaw, except in the dream the bundle in her arms was empty.

He woke up with a start when the truck stopped. Late afternoon now, the sun is low. Through the gap in the canvas he could see a gate. A large wooden gate in a fence that stretched in both directions. Above the gate was a sign in German.

"Stalag VIII-A," a guard announced. "End of journey."

The prisoners climbed down, stiff and sore. Dick's legs nearly buckled. He leaned against the truck until his balance returned.

The camp spread out before them. Rows of wooden barracks, all identical, all grey. Guard towers at regular intervals. Two fences, one inside the other, with space between. Warning signs on the inner fence. Beyond the camp, flat fields stretching to the horizon.

German soldiers with clipboards began processing the new arrivals, checking names against lists, assigning barracks numbers. The prisoners stood in line, their faces blank with exhaustion.

When Dick's turn came, he gave his name and rank. The German clerk wrote it down without looking up.

"Barracks fourteen," the clerk said. "Follow that path."

Dick nodded and moved on. Tom was assigned to the same barracks. Walsh was sent to barracks fifteen, just one over. They said brief goodbyes.

Barracks fourteen was identical to all the others. Long and narrow, with two rows of wooden bunks. A stove at one end, unlit. Windows too high to see out of properly. The smell of unwashed men and old wood.

A British sergeant who'd been there longer showed Dick and Tom to empty bunks halfway down the building. "Lower one's better," he said. "Warmer in winter. Roll call's at five in the morning."

Dick claimed the lower bunk and sat down on the thin mattress. First time he'd had a space that was his own since leaving Crete.

Tom took the bunk above and immediately lay down with a groan. "Christ, I'm tired."

"Yeah."

"You think they'll feed us tonight?"

"Don't know."

Tom was quiet for a moment. Then he said, "That city. Warsaw. That was hard to see."

"Yeah."

"You reckon Newcastle's all right?"

"I hope so."

"But you don't know."

"No."

Tom shifted on the bunk above. Then he said, "Betty. I keep thinking about her. About whether she's safe. About whether she'll wait."

Dick pulled out the photograph and looked at it again. His family. He thought about Eileen anyway. About her voice on the telephone exchange. About her face in the park.

"They'll wait," Dick said.

"You sure?"

"No. But we have to believe they will."

Tom didn't answer. Dick put the photograph away and lay back on the bunk. The mattress was thin and the boards under it were hard, but after days in trucks and trains it felt like luxury.

He closed his eyes and tried not to think about the woman in Warsaw. Tried not to think about how far he was from home.

But the thoughts came anyway. They always did.

He pulled the thin blanket up and closed his eyes. Tomorrow would be roll call and work details and whatever came next. But tonight he carried Warsaw with him. The woman's face. Her steady gaze. He would carry it forward, wherever this led.

Eventually exhaustion won and he slept.

CHAPTER 10 — STALAG LIFE

The whistle cut through Dick's sleep. He jerked awake. Darkness. The smell of unwashed men. Cold pressing through gaps in timber walls.

"Up! Raus! Schnell!"

Guards moved down the aisle between bunks, rifle butts banging against wooden frames. Men groaned and stirred. Above Dick, Tom's bunk creaked as he swung his legs over.

"Christ," Tom muttered. "What time is it?"

"Too bloody early," someone answered.

Dick sat up. The cold hit him. His chest tightened. His breath came out in white clouds. The stove at the far end was dead, just cold iron and ash. Outside the high windows, the sky was still dark.

He pulled on his boots, his fingers clumsy and stiff. The leather was rigid and cold. His feet had blisters on top of blisters. Every morning he wondered if this would be the day his feet gave up. Every morning they kept working.

"Five minutes!" a guard shouted. "Outside for roll call!"

The barracks erupted. Forty men trying to dress, find

boots, relieve themselves in the bucket by the door. The British sergeant who'd shown Dick and Tom to their bunks the night before, a man named Fletcher, moved through the chaos with the calm of long practice.

"Stay in your sections," Fletcher said. "They count by fives. Know who's next to you. If the count's off, they start again and we're all out there longer."

Dick nodded and found Tom. They moved toward the door together.

Outside was worse. The cold went through Dick's jacket. Frost covered the ground, white and crystalline in the few lights. His breath froze in his throat.

The prisoners formed rough lines, five across, multiple rows deep. Guards walked along with clipboards, counting in German. Dick stood between Tom and a British corporal named Davies who'd been captured at Dunkirk.

Dick stamped his feet, trying to keep blood moving. Around him, other men did the same. Some coughed, deep rattling sounds that didn't stop. Some just stood with arms wrapped around themselves, shivering.

The counting took twenty minutes. A German officer checked numbers against his list, nodded to the guards, and spoke briefly.

"Barracks fourteen, return for breakfast."

Dick shuffled back with the others. His feet were numb. His hands were numb.

Inside felt fractionally warmer. Someone had got the stove going. Men gathered around it, holding their hands out.

Breakfast arrived in a metal container carried by two prisoners. A guard watched as Fletcher ladled contents into bowls. Dick had been issued a metal bowl and spoon the night before. He held them now and waited.

When he reached the front, Fletcher poured thin grey liquid into Dick's bowl. "Soup," Fletcher said, though the word was generous. It looked like dishwater and smelled of nothing.

Dick took his bowl and found a spot on his bunk. Tom sat beside him, looking at his portion with an expression between disgust and despair.

"This is breakfast?" Tom asked.

"This is breakfast."

"Christ."

Dick lifted the bowl to his lips and drank. Lukewarm, tasting of dirt and something vaguely vegetable. He forced himself to swallow, to put something in his stomach.

Tom did the same, wincing. "My mum's porridge was never this bad."

"Nothing was ever this bad."

They sat in silence, drinking their soup, making it last because there would be nothing else until midday. Around them, other men did the same. When the bowl was empty, Dick licked it clean. Everyone did. Every drop mattered.

Fletcher came by and sat on the bunk across from them. Maybe thirty-five, with a thin face and grey in his hair that probably hadn't been there before the war. He'd been at Dunkirk, captured when his unit was cut off. A prisoner for a year and a half now.

"First night all right?" Fletcher asked.

"Barely slept," Tom said.

"You'll get used to it. It takes a week or two. The body adapts."

"What happens now?" Dick asked. "After breakfast?"

"Work detail. They'll call names. You go where they send you. Could be anything. Depends on what needs doing."

"How long?"

"Till they tell you to stop. Could be hours. Could be all day."

Dick nodded slowly.

"Stick together if you can," Fletcher said, looking at Dick and Tom. "Australians with Australians. Watch each other. That's how you survive this."

"What about food?" Tom asked. "Is it always like this?"

Fletcher's expression didn't change. "It's better some days. Worse on some others. Red Cross parcels come through sometimes. Not often. You'll learn to make things last."

A whistle blew outside. Fletcher stood. "That's for work detail. Come on."

They filed outside again. The sun was up now, just a pale disc behind thick clouds. The temperature had risen maybe a degree or two.

A German sergeant stood with a clipboard, calling out names and assigning work parties. Dick heard his own name and Tom's called together, along with a dozen others. They were directed to a truck waiting by the gate.

"Where are we going?" Dick asked one of the guards as they climbed into the back.

"Holz," the guard said. "Wood. You carry wood."

The truck drove for twenty minutes through flat countryside, all grey and brown, fields bare, trees skeletal. They stopped at a logging area where huge piles of cut timber sat waiting.

Dick grabbed a log and hefted it. The wood was green and heavy, his hands still numb. He carried it to the truck and heaved it up. Then went back for another.

Tom worked beside him. Bend, lift, carry, load. Over and over. The work warmed them gradually though the cold never left. Dick's hands began to hurt as feeling returned. His back began to ache. His blistered feet screamed with every step.

But he kept moving. Stopping meant drawing attention. Attention meant punishment.

They worked through the morning without break. One of the civilian workers, an older man with a grey beard, watched them. He didn't speak, but once, when a guard wasn't looking, he set a chunk of dark bread on a stump where Dick would see it. Dick palmed it when he passed and shoved it in his pocket. The civilian looked away.

At midday they were given water and told to sit for fifteen minutes. Dick collapsed against the truck wheel. He pulled out the bread and split it with Tom. Hard as stone, but it was food. They chewed, making it last.

"How long do you think this war's going to last?" Tom

asked.

Dick thought about Warsaw. About seeing what the war had already done. "I don't know. Could be years."

"Years."

"Could be."

Tom was quiet. Then he said, "Betty won't wait years. No girl will."

Dick thought about Eileen. About her offer to get engaged. About his refusal. He'd written three letters from the transit camp. No idea if she'd received them.

"Some will wait," Dick said, though he wasn't sure he believed it.

The break ended. They went back to work. The afternoon stretched on. Dick's mind drifted while his body moved. Newcastle. The beach. His mother's kitchen. His sisters arguing about nothing important.

They returned to camp as the sun was setting. Dick's body was past exhaustion. They filed into the barracks and found Fletcher ladling out the evening meal. Soup again, this time with actual pieces of turnip. And bread, a single slice.

Dick took his portion and sat on his bunk. He ate the bread first, chewing each bite, making it last. The soup he drank, tipping the bowl to get every drop. When he was finished, his stomach was still empty. But less empty than before.

Tom climbed into the bunk above and lay down with a groan. "I can't do this every day."

"You will," Dick said. "Because there's no choice."

"Bloody hell."

Dick lay back and closed his eyes. Around him, the barracks settled. The men talked quietly. Someone wrote a letter. Someone mended a shirt. The stove crackled.

The days fell into pattern. Roll call at dawn. Thin soup. Work detail. Different jobs on different days. Hauling coal. Digging drainage ditches. Loading trucks. Always under guard, always watched, always cold and hungry.

Dick learned which guards were dangerous. Hartmann, the sergeant who ran roll call, was strict but fair. Schmidt, who supervised some work details, would look the other way if prisoners rested. But Kruger, a young corporal with cold eyes, would hit a man for looking at him wrong.

In late October, Dick was on a work detail repairing a bombed building when Miller, working beside him, suddenly stopped moving. His shovel fell from his hands. He stared at his fingers.

"Miller?" Dick said.

Miller held up his right hand. The fingertips were white, waxy-looking. No feeling in them.

"Frostbite," one of the other prisoners said. "Get him inside. Now."

They got Miller to a shed and rubbed his hands, trying to get circulation back. The guard watched but didn't stop them. When feeling returned, Miller gasped and doubled over. The pain, someone said, was worse than the numbness.

That night, Dick looked at his own hands. Red and

chapped, but no white patches. Not yet. He tucked them under his armpits for warmth.

Fletcher found him by the stove. "You need to be more careful. Winter here isn't like Australia."

"I know."

"No, you don't. Not yet. But you will." Fletcher looked at Dick's hands. "Wrap them when you can. Keep them dry. Frostbite will take your fingers faster than you think. I've seen it happen."

Dick nodded. That night he tore strips from his spare shirt and wrapped his hands before sleeping.

November came. The ground froze solid. Snow fell and didn't melt. The stove in the barracks burned constantly but never warmed the building properly. Men slept in all their clothes, two or three to a bunk, sharing body heat.

One morning in mid-November, word went through the barracks: Red Cross parcel.

The men gathered after the evening roll call. A single parcel for forty men. Fletcher divided it carefully. Dick got a quarter tin of meat, one square of chocolate, and two cigarettes.

He ate the chocolate immediately, letting it melt on his tongue. The richness hit his empty stomach hard, but it was worth it. The cigarettes he saved. They could be traded.

Tom got similar portions. He smoked one cigarette immediately, savoring each drag. The other he put away.

"Remember real food?" Tom asked, smoke trailing from

his mouth.

"Yeah."

"My mum's roast. Potatoes. Gravy. Pavlova after." Tom closed his eyes. "I'd kill for pavlova right now."

Dick thought about his mother's cooking. About the kitchen smell of dinner. About his sisters setting the table. About sitting down together as a family.

"We'll have it again," Dick said. "When we get home."

"You sure?"

"No."

They sat with that honesty.

Three weeks after arriving at the camp, mail call was announced after the evening roll call. The men gathered in the parade ground, breath steaming, as a German corporal read names from a stack of letters.

Dick stood frozen as names were called. He hadn't expected anything. Mail took months to catch up with prisoners. But the guard kept reading names, and then Dick heard it.

"Roberts, James."

For a moment Dick couldn't move. Then he stepped forward, his hand raised. "Here."

The corporal checked his list, nodded, and handed over a single thin envelope. Dick took it, his hands shaking. The envelope was creased and dirty, stamped multiple times, addressed in his mother's careful handwriting. Broadmeadow, Newcastle.

He stepped back into the crowd. Tom's name wasn't called. Neither was Miller's. Most men got nothing. But

Dick had this thin piece of paper that had travelled halfway around the world.

He waited until he was back in the barracks, until he'd found a spot on his bunk away from the others, before he opened it. He unfolded the thin paper.

Dear Dick,

We received word that you were captured on Crete and are a prisoner of war in Germany. The Red Cross sent a telegram. Your father and I were so relieved to know you're alive. We had feared the worst when we heard the reports from Crete.

Everyone here is well. The house is the same as you left it, though quieter without you. Jean and Lorna have taken on more responsibility, and I don't know what I'd do without them. Jean got a job at the telephone exchange. Eileen works there, as you know. She helped Jean get the position and still asks after you when I see her at the shops.

Dorrie is doing well at school. She's very practical, like you. Always helping with the younger ones. Shirley and Meryl miss you terribly. Meryl asked when you're coming home and I didn't know what to tell her. I said soon, but I don't know if that's true. I pray it's true.

The garden is doing well. Your father planted tomatoes and they came up better than expected. We've been making preserves for winter. It seems silly to write about tomatoes when you're so far away, but I want you to know that life continues here. We're waiting for you.

We're sending parcels through the Red Cross when we can. I don't know if they'll reach you, but we're trying. Your father works longer hours at the paper now.

Everyone works longer hours. But we manage.

Please write when you can. We worry. I worry. I know you're strong. I know you'll come through this. But I'm your mother, and I worry anyway.

All my love, Mum

Dick read the letter twice, then a third time. Then he folded it and held it against his chest, his eyes closed. His mother's words. His mother's handwriting. Proof that Newcastle existed, that the world he'd come from hadn't been a dream.

Jean at the telephone exchange. Eileen helped her get the job. Eileen is still asking after him. She hadn't forgotten. Hadn't moved on.

Meryl asked when he was coming home. His baby sister, ten years old now, not understanding why her brother had been gone so long.

He pulled out the photograph and looked at it alongside the letter. His mother, his father, his five sisters. They were there, in Newcastle, living their lives. Growing tomatoes. Making preserves. Waiting for him.

Dick found paper and a stub of pencil he'd traded for. The camp allowed prisoners to send one letter per week, a single sheet, censored before being mailed.

Dear Mum,

I received your letter. Thank you. It meant more than I can say.

I'm well. The camp is adequate. We get food and shelter. Don't worry about me. I'm managing fine.

Tell the girls I think about them every day. Tell Meryl

I'll be home when the war ends. I don't know when that will be, but it will happen. Tell Jean I'm glad she's working. Tell Lorna not to argue with everyone too much.

Thank Eileen for helping Jean. Tell her I'm thinking about her. I wrote to her from the transit camp, but I don't know if my letters got through. If she's written to me, I haven't received anything yet. Mail takes time.

The days here are long but bearable. I'm with other Australians, which helps. Tom from home is here, in the same barracks. We look after each other. We'll get through this.

I'm sorry I can't write more. They limit what we can say and how much. But know that I'm all right. Know that I'm doing what I have to do. Know that I think about home every day.

All my love, Dick

He folded the letter and gave it to Fletcher to add to the stack for the censors. It would take weeks, maybe months, to reach Newcastle. But his mother would open it and know he was alive and thinking of her.

That night, lying in his bunk with the letter from home pressed against his chest under his shirt, Dick felt hope. Not for release, not for freedom. Just hope that he would survive. That his family would survive. That when this was over, there would still be something to go home to.

Above him, Tom was quiet. "You get a letter?" Tom asked after a while.

"From my mother."

"What'd she say?"

"Everyone's well. Life goes on. Jean's working. Eileen asks after me."

"That's good. That she asks after you."

"Yeah."

"I wish Betty would write. I've sent four letters. Haven't heard anything back."

"Mail takes time."

"Yeah." Tom didn't sound convinced.

December came. Dick marked it by scratching lines on the wooden post beside his bunk. Other men did the same, counting days, trying to hold onto time in a place where time had lost its shape.

Christmas approached. Fletcher organized what he could. The prisoners pooled resources. Someone had saved chocolate from a Red Cross parcel. Someone else had hoarded dried fruit. They made something almost like pudding.

On Christmas Eve, the Australians gathered in their section. Maybe thirty of them in the camp, scattered across different barracks but united by distance from home. They sang carols, voices thin and cracked but real. Silent Night. O Come All Ye Faithful.

Dick thought about last Christmas. The house in Alfred Street, warm and bright. His mother cooking, his sisters laughing. Church and dinner. It felt like a lifetime ago.

"What do you think they're doing now?" Tom asked. It was late, the barracks mostly quiet. "Back home. On Christmas."

Dick looked at the darkness beyond the windows. It

was afternoon in Newcastle now, the height of summer. "Going to the beach, probably. Having a barbecue. My mother makes a trifle every year."

"My mum does roast lamb. Always lamb. And pavlova after. Betty usually comes over for dinner." Tom paused. "I wonder if she'll go this year."

"Maybe."

"Or maybe she's moved on."

"Don't think like that."

"Why not? It's possible. We've been gone for eight months. She's nineteen. She's not going to wait forever."

Dick didn't answer. Because Tom was right. It was possible.

"Eileen hasn't written," Dick said quietly. "Not since before I was captured. I don't know if that means she hasn't written or if the letters just haven't caught up."

"But your mum said she asks after you. That means something."

"Does it? Or is she just being polite?"

They sat with that question. Outside, snow was falling. Inside, the stove burned low. Men slept, or pretended to sleep.

On Christmas morning, the Germans gave the prisoners an extra portion of bread and a thin slice of sausage. The best meal Dick had eaten in months. He ate, making it last, trying to taste each bite.

At noon, mail call was announced. More names are called. More letters handed out. Dick's name wasn't among them.

But Tom's name was called. He stepped forward and received two letters. One from his mother. One with Eileen's handwriting on the envelope.

Tom brought the letters back, looked at the second one, and handed it to Dick without a word.

Dick took it. The envelope was addressed to Tom, but Eileen must have included something for Dick inside. He watched Tom open his letter from his mother while Dick opened the one from Eileen.

Inside was a single sheet folded around a second, smaller piece. The smaller piece had Dick's name on it. Tom handed it over.

Dick unfolded it. Eileen's handwriting, neat and precise.

Dear Dick,

I'm writing to Tom because I don't know your address. I hope he'll pass this along to you. I hope you're all right. I hope you're not hurt.

Your mother told my mother that you were captured. I went to see her right away. She showed me the telegram from the Red Cross. I don't know what to say except that I'm glad you're alive.

I got your letters from the transit camp. All three of them. They took a while to reach me, and by the time they did, you'd already been moved. I wrote back but I think my letters are chasing you around Europe. I'll try again with this one.

You told me to get on with my life. You said you didn't want me to wait. But I don't know how to do that. I don't know how to stop thinking about you. I don't know how to stop hoping you'll come home.

Jean works with me now at the exchange. She's good at it. Your mother seems to be managing, though I know she worries. Everyone worries.

I won't make any promises. You didn't want promises. But I'm writing to you anyway. I'm thinking about you. I'm hoping that wherever you are, you're finding a way to survive.

Please write if you can.

Eileen

Dick read it twice, then folded it and put it in his pocket with the photograph and his mother's letter. His collection of paper, his connection to home.

She'd written. She was thinking about him. She hadn't moved on.

"Good news?" Tom asked.

"She wrote. She's thinking about me."

"That's something."

"Yeah." Dick looked at Tom. "What about you? What'd your mother say?"

Tom's face had gone pale. He was staring at the letter in his hands. "Betty's engaged. To some bloke from Merewether. They're getting married in February."

"Tom, I'm sorry."

"She didn't even write to tell me herself. My mother told me. She thought I should know before it happened." Tom's voice was flat, empty. "Seven months. That's how long she waited. Seven bloody months."

"Tom."

"Don't." Tom folded the letter and shoved it under his pillow. "Don't say it's all right. Don't say she has a right to move on. I know all that. I just don't want to hear it right now."

He rolled onto his side, facing the wall, his back to Dick. Dick wanted to say something comforting. But there was nothing. Tom had lost the girl he loved. All Dick could do was sit there and be present.

That night, Dick pulled out Eileen's letter and read it again by the dim light from the stove. I don't know how to stop thinking about you. Those words meant something. They had to.

He thought about Tom, about Betty's engagement, about how quickly a person could be replaced. Seven months. And Dick had been gone the same amount of time. Would Eileen wait longer?

January came, marked by more scratches on the post beside Dick's bunk. The cold was merciless.

One morning, the Australian from Melbourne, a man Dick had spoken to a few times, didn't wake up for roll call. The guards checked him. His body had simply given up during the night.

They buried him that afternoon in the frozen ground. A chaplain said words over the hastily dug grave. Dick stood with the others in the snow and understood that this could be any of them. Death took who it wanted, when it wanted.

He wrote to Eileen that week.

Dear Eileen,

I got your letter. Thank you for writing. Thank you for

not giving up on me.

I can't promise when I'll be home. I can't promise anything except that I'm trying to survive this. That's all any of us can do.

Tom got a letter from his mother. Betty's engaged to someone else. It's hard watching him deal with that. It makes me think about us, about what you said in the park before I left. Maybe you were right. Maybe I should have said yes.

But I can't think that way now. I can only think about getting through each day and hoping that when this is over, there's still something to go home to.

Please keep writing if you can. Your letters matter more than I can say.

Dick

He gave the letter to Fletcher for the censors. It would take months to reach her. Her response, if she sent one, would take months to reach him.

February came. The cold began to ease slightly. Instead of deadly it became merely brutal.

March arrived. Dick had been a prisoner for five months. His body had shrunk, adapted to less food. His mind had adapted to the routine. But he held onto his mother's letter. He held onto Eileen's letter. He held onto the photograph. These were proof that another life existed, that he was more than just a prisoner.

At night, lying in his bunk, he would close his eyes and imagine Newcastle. The beach. The sound of surf. His mother's kitchen. Eileen's voice. He held these images like treasures, refusing to let them fade.

Because if he forgot who he'd been, if he forgot what he was fighting to return to, then the camp had won. And Dick wasn't ready to let that happen.

The days continued. The war continued. And Dick continued with them, one day at a time, waiting for the moment when he could finally go home.

CHAPTER 11 — THE WORK BRIGADE

The names were called at morning roll call on a day in late March when frost still silvered the ground but the air held a hint of warmth beneath the cold. Dick stood in his usual spot, fifth row back, with Tom to his right and Miller to his left. His breath came out in white clouds. His feet were numb in his boots.

The German sergeant, Hartmann, read from a clipboard. Not the usual count by fives, but specific names. Dick's attention sharpened.

"Roberts, James."

Dick's heart kicked. He raised his hand. "Here."

"Wilson, Thomas."

"Here," Tom said.

More names. Miller. Walsh from the next barracks. A dozen Australians and New Zealanders. Two British soldiers. Hartmann finished reading and looked up.

"These men will gather belongings and report to the main gate in one hour. Work brigade assignment. Move quickly."

Dick glanced at Tom. Tom's face was carefully blank.

They filed back to the barracks with the others. Fletcher met them at the door.

"Work brigade's usually farm labour," Fletcher said. "Smaller camps, less structure. Food can be better or worse depending on the farmers. Guards are often older men, reservists."

"How long?" Dick asked.

"Could be weeks. Could be months. Could be the rest of the war." Fletcher looked at each of them. "Watch yourselves. Smaller camps mean less oversight."

Dick climbed to his bunk and gathered what he had. A spare shirt he'd traded for. A wooden spoon carved by a New Zealander who'd since died. The letters from his mother and Eileen, folded so many times the creases were wearing through. The photograph, its edges soft from handling.

He tucked everything into his pockets. What you carried had to fit on your body or be left behind.

Tom dropped down from the upper bunk. "Ready?"

"As I'll ever be."

They walked together to the main gate. The others were already gathering there, fifteen men total, stamping their feet, breath steaming. Guards checked names against the list. When everyone was accounted for, the gate opened and they were marched out through the wire.

Dick looked back once at Stalag VIII-A. Five months there. He'd learned the rhythms, the dangers, the small mercies. Now he was leaving.

They walked for hours. The road was frozen hard,

rutted from farm carts. The countryside opened around them, flat fields stretching to low hills, bare trees like sentinels. The sky was grey and close.

The guards marching with them were older men. One was perhaps fifty, with grey hair showing under his helmet and a pronounced limp. The other was younger, maybe thirty, with a thin face and nervous eyes that constantly scanned the prisoners.

Dick walked beside Tom. His feet hurt. The blisters from Crete had healed into calluses, but new blisters were forming over the old scars. He focused on putting one foot in front of the other.

"You reckon this is better or worse?" Tom asked quietly.

"Won't know till we get there."

"Helpful, mate. Really helpful."

"You want me to lie?"

"Wouldn't mind."

They walked on. Around midday, the guards called a halt. Dick sat on a stone by the roadside and pulled out a piece of bread he'd saved from breakfast. He ate half and put the rest back.

The older guard with the limp walked past, pulled out a pipe, and started filling it with tobacco.

"Where we going?" Tom asked in German.

The guard looked at him, then lit his pipe. "Farm. Ten kilometers more. You work, you eat. Simple."

"What kind of work?"

"Farm work. What else?" The guard's accent was thick but his English was clear. He looked at the prisoners.

"War is stupid. You know this. I know this. But here we are."

He walked on, smoke trailing behind him.

"Seems all right, that one," Tom said.

"Maybe."

They reached the farm as the sun was setting. It sat in a shallow valley beside a creek, the water running dark between banks of frozen mud. A main house, solid and square. Two barns. Several sheds. Fields stretching in all directions, all of them brown and bare.

The guards directed them to one of the sheds. Rough timber with a dirt floor and high windows. A stove in one corner, unlit. Wooden pallets lined the walls.

"Schlafen," the older guard said, pointing at the pallets. Sleep.

The prisoners filed in and claimed spots. Dick and Tom took pallets near the stove. Walsh settled nearby, along with Miller and the two British soldiers, Private Cooper and Corporal Andrews.

When the guards had left and locked the door, the prisoners sat in the gloom.

"Well," Miller said. "At least it's not the main camp."

"At least there's that," Walsh agreed.

Dick pulled the photograph from his pocket and looked at it by the dim light. His family. His mother's gentle smile. His father's stern expression. His five sisters. He wondered what they were doing. It would be morning in Newcastle, autumn turning toward winter. His mother making breakfast. His sisters getting ready for school or work.

"Your family?" Cooper asked.

"Yeah."

"Big lot."

"Five sisters."

"Christ. Rather you than me, mate."

Dick put the photograph away. "You got family?"

"Wife and a daughter. Three years old when I left. She'll be five now. Won't even remember me." Cooper's voice was flat. "Wife writes when she can. Says our girl asks about the man in the photograph."

No one knew what to say. They sat in silence.

The door opened and the older guard came in carrying an armload of wood. He knelt with difficulty because of his leg and began building a fire. When it was lit and catching, he stood and looked at the prisoners.

"My name is Werner," he said in careful English. "This is my farm. You work for me now. The other guard is Beck. Young, nervous. Do not make him nervous. Understand?"

The prisoners nodded.

Werner looked at each of them. "I do not care about politics. I do not care about the war. I care about the farm. You work well, you eat well. You cause trouble, you go back to the main camp. Simple rules."

He walked to the door, then paused. "I was in the last war. France. I know what it is to be a prisoner. I will not be cruel. But I will not be soft. Do we understand each other?"

"Yes," Dick said. The others murmured agreement.

Werner nodded and left, locking the door.

The fire crackled and slowly warmed the shed. The prisoners moved closer, holding their hands out.

"Seems decent," Tom said quietly.

"Maybe. We'll see."

The next morning began with a whistle at dawn. The door opened and Beck stood there with his rifle. "Raus! Schnell!"

They filed out into the cold. Werner was waiting in the yard, along with a civilian. A man in his forties with a weathered face and the build of someone who'd worked hard his entire life. This was Heinrich, the farm manager. He looked the prisoners over with the practical eye of someone assessing tools.

"Can any of you plough?" Heinrich asked in German. Werner translated.

Dick and two others raised their hands.

"Good. You three with me. The rest will repair fences and clear drainage ditches."

They were split into groups. Dick found himself with Miller and a New Zealander named Price, following Heinrich toward the fields. The ground was still partially frozen, but the frost was breaking up.

Heinrich showed them the plough, an old wooden thing pulled by two massive grey horses. He demonstrated the technique, his hands sure on the handles.

"Keep the furrow straight. Do not go too deep. The soil here is thin. Work steady, not fast. The horses know what to do."

Dick took the handles first. The horses waited while he positioned himself. Heinrich clicked his tongue and they moved forward, the plough biting into earth, turning it over in a dark wave.

It was hard work. The plough wanted to twist and buck. Frozen patches caught and jarred his arms. His shoulders ached within an hour. But there was rhythm to it, the simple repetition, walking behind the horses, the smell of turned earth.

They worked through the morning. Heinrich watched from the edge of the field, smoking and occasionally calling out corrections. At midday, Werner brought food. Not thin soup, but actual food. Thick bread with butter. Cheese. Sausage. A thermos of hot coffee.

The prisoners sat in the turned earth and ate. Dick hadn't tasted butter in months. The richness almost made him sick, but he forced himself to eat slowly. His body needed this.

"This is better than the main camp," Miller said through a mouthful of bread.

"Don't get used to it," Price replied.

But it didn't change. The days fell into a pattern. Work from dawn to dusk, hard labour that left them exhausted but somehow more alive than sitting in barracks at Stalag VIII-A. The food was better, not generous but adequate. Real food. The guards watched but didn't harass. Werner treated them like workers, not prisoners. Even Beck, though nervous and quick to bark orders, didn't use his rifle except to gesture.

Dick ploughed fields, repaired fences, cleared stones, dug drainage ditches. His hands hardened with new calluses. His body grew stronger from the work and better food.

In the evenings, after work was done and they were locked in the shed, the prisoners talked. Not about the war or when they'd go home. They talked about small things. The work they'd done that day. The horses and their personalities. The weather.

Dick wrote letters when he could get paper from Werner. He wrote to his mother, careful words that said little but meant everything. He wrote to Eileen, longer letters full of things he shouldn't say but couldn't help saying.

Dear Eileen,

I'm on a work brigade now. A farm. It's better than the main camp. The work is hard but honest. I can see the seasons changing. Spring is coming.

I think about you more than I should. I think about that evening in the park, about what you asked and what I said. I think I was wrong. Maybe making promises would have been better than making none.

But I can't think that way. I can only think about getting through each day and hoping you're still there when I get home. If I get home.

Please keep writing. Your letters are all I have.

Dick

He gave the letters to Werner to mail. Weeks later, a letter came back from his mother. She wrote about the house, about his sisters, about small things that meant everything. Jean was engaged now, to a man who worked at the steelworks. Lorna had got a job at a shop in town. Dorrie was doing well at school. Shirley and Meryl missed him terribly.

At the end, almost as an afterthought: Eileen asks after you whenever I see her. She's well. She's waiting, I think. Though she doesn't say it directly.

Dick read that line five times. She's waiting. Maybe.

Tom received a letter too, from his sister. No mention of Betty. Tom read it once, folded it carefully, and didn't speak for the rest of the evening.

April came. Dick noticed it first in the light, which came earlier and stayed later. Then in the softness of the ground underfoot, no longer frozen. Then in the fields themselves, which turned from brown to green in the space of a week.

They planted potatoes, working in long rows, dropping seed potatoes into furrows and covering them with earth. The work was backbreaking but there was hope in it. Putting things in the ground, trusting they would grow.

One afternoon, while Dick was working near the barn, a woman came out of the farmhouse. She was perhaps forty, with blonde hair going grey and a face weathered by sun and work. She carried a wooden bucket filled with fresh milk and cups.

She approached cautiously, glancing back at the house. Werner saw her but turned away.

"Trinken," she said softly, offering the cups. Drink.

Dick took a cup and drank. The milk was still warm, rich

and slightly sweet. The best thing he'd tasted in a year.

"Danke," Dick said.

The woman nodded. She looked at him for a moment, then returned to the house without another word.

The next day she came again with milk. And the day after. It became routine.

But Beck was changing. Dick noticed it first in the way he watched them work. His eyes were harder, his jaw tighter. He stood with his rifle ready, finger too close to the trigger. He shouted more, longer, about smaller things.

One morning, Beck kicked over a bucket of water Price had just filled. "Too slow!" he shouted, though Price hadn't been slow at all.

Another day, he made Walsh stand at attention in the yard for an hour because Walsh had coughed during roll call.

Werner watched but said nothing. The prisoners moved more carefully, making every action visible and unthreatening.

"What's wrong with him?" Tom asked one evening.

"Don't know," Dick said. "But stay clear of him."

The next morning, Heinrich appeared beside Dick while he was repairing a fence post. Heinrich glanced at Beck, then leaned close and spoke in low German.

"Be careful with that one. He received news yesterday. His brother was killed in Russia. He blames everyone now."

Dick nodded his thanks.

That explained it. Beck's brother, dead somewhere in Russian snow. And here were enemy prisoners, alive and working, eating German food.

Dick understood. But understanding didn't make Beck less dangerous.

The next day Beck was worse. He prowled the edges of the work areas like a predator. The prisoners moved slowly, carefully.

Two days later, Walsh was repairing a fence when Beck appeared behind him.

"You're going too slow," Beck said.

"No sir, I'm working steadily."

"You're sabotaging! I see it!"

"I'm not, I swear."

Beck raised his rifle butt and brought it down across Walsh's shoulders. Walsh went down hard. Beck kicked him once, twice.

Werner came running, his bad leg making him lurch. "Beck! Stop!"

Beck turned on him. "He was sabotaging!"

"He was working. I was watching." Werner stepped between Beck and Walsh. "Go to the house. Now."

For a moment Dick thought Beck would shoot Werner. Beck's face twisted, his finger on the trigger.

Then Beck lowered the rifle and walked away toward the house.

Werner helped Walsh up. Walsh's face was already swelling, his lip split. He spat blood.

"Get him to the shed," Werner said to Dick. "I will bring water."

That night, Dick and Tom cleaned Walsh's face while the others watched in silence. Walsh's ribs were bruised, maybe cracked. His face would be black and blue by morning.

"I wasn't going slow," Walsh said. "I wasn't sabotaging."

"We know," Dick said.

Werner came to the shed after dark. He looked at Walsh, then at all of them.

"I apologize. Beck is frightened. The war is not going well for Germany. He takes this fear out on you. I will try to control him. But be careful. Do not give him reason."

After Werner left, they sat in the darkness.

"He's going to kill someone," Walsh said quietly.

"Maybe," Dick replied.

The next three days Beck stayed near the house. They saw him watching from windows, but he didn't come out to supervise. Werner handled the work details alone.

On the fourth day Beck returned. He was quieter but his eyes were still hard. He watched them work without speaking. The prisoners gave him a wide berth.

Then, on a Tuesday morning in early May, everything changed again.

Dick was ploughing with the grey horses when he heard shouting from the barn. He looked up to see Beck dragging Price across the yard by his collar. Price was protesting, his voice high with fear.

Werner came out of the house, moving as fast as his bad leg would allow. "Beck! What is this?"

"He was stealing!" Beck's face was red, his voice shaking. "I saw him! Taking eggs from the henhouse!"

Price was shaking his head. "I wasn't, I swear. I was just looking. I didn't take anything."

"Liar!" Beck raised his rifle, bringing the butt down toward Price's head.

Werner grabbed the rifle, his strength surprising. "Stop! This is my farm! My authority!"

Beck's face twisted with rage and grief. His finger was on the trigger. One twitch and Werner would be dead.

Then Beck seemed to collapse. He lowered the rifle, released Price, and walked away toward the road. They watched him go.

Werner looked at Price. "Did you take eggs?"

"No sir. I was just looking at the chickens."

Werner nodded slowly. "I believe you. But stay away from the henhouse." He looked at all the prisoners. "Everyone must be very careful now."

That evening, Beck didn't return. Werner stood guard himself, sitting by the shed with his rifle across his knees. He looked older, worn down.

Dick approached him carefully. "What will happen to Beck?"

Werner shrugged. "He will come back. Or he will not. The war makes men strange. He has lost his brother. He sees you alive and blames you. It is not rational. But grief is not rational."

"Will he come back?"

"Probably. He is a soldier. He will do his duty." Werner's pipe had gone out. He didn't bother to relight it. "I am tired of this war. But I will do my duty until it ends. You will do yours. That is all any of us can do."

Beck returned three days later. Quieter, but his eyes were still watchful. He and Werner barely spoke. The prisoners gave him a wide berth and worked with exaggerated care.

The days continued. May deepened. The fields grew green and thick. The potatoes they'd planted sprouted and grew. The work continued, endless but rhythmic.

Dick received another letter, this time from Eileen. It had taken three months to reach him.

Dear Dick,

Your mother gave me your new address. A work brigade, she said. I hope the work isn't too hard.

I got your last letter. The one about promises. You were right, I think, not to make promises. But you were also wrong. Because we made promises anyway, didn't we? Just by caring. Just by writing letters. The promises were already there, even if we didn't say them out loud.

I'm still at the exchange. Jean works with me now. We talk about you sometimes. She misses you. Your whole family misses you.

I don't know what else to say except that I'm here. I'm waiting. Not because you asked me to, but because I can't seem to do anything else. Does that make me foolish? Maybe. But there it is.

Please keep writing. Please come home.

Eileen

Dick read the letter until the words blurred. I'm here. I'm waiting.

He showed it to Tom, who read it and handed it back without comment. Tom hadn't received anything from Betty since the engagement announcement.

"You're lucky," Tom said finally. "She's still there. Don't take that for granted."

"I won't."

"Some of us don't get that."

Dick folded the letter and put it with the others. He thought about Eileen at the exchange, her voice calm and precise as she connected calls. He thought about her hands, her face, the way she'd looked at him in the park.

That night, lying on his pallet with the others sleeping around him, Dick pulled out the photograph and looked at it by moonlight. His family. Eileen wasn't in the picture, but she was part of the world he was fighting to return to. She and his family and Newcastle and everything that had been his life before.

He would survive this. He would get through the work and the hunger and the danger and the endless waiting. Because Eileen was waiting. Because his family was waiting.

The summer stretched ahead, full of work and heat and the slow turning of seasons. The war continued somewhere beyond the farm. But here, in this small pocket of Poland, Dick ploughed fields and planted crops and received letters from home.

He survived one day at a time, holding onto the threads that connected him to the life he'd left behind. And he waited. Just as Eileen was waiting. All of them separated by distance and war and time, but still connected.

He closed his eyes. Tomorrow would bring more work. More uncertainty. But tomorrow would also bring him one day closer to the end.

For now, he held onto that. The photograph in his pocket. The letters from home. The knowledge that somewhere, people were waiting for him to return.

That would have to be enough.

CHAPTER 12 — THE FIRST ESCAPE

The snow started on a Tuesday in late November, fat flakes that fell straight down through still air. Five months at the farm had passed in the rhythm of seasons and work: summer heat giving way to autumn cold, then this first real snow. By evening, the fields were white and the temperature had dropped enough to make breathing hurt. Dick stood in the doorway of the shed watching it fall, his arms wrapped around himself against the cold.

Behind him, the others were gathered around the stove, trying to extract what little warmth it offered. Tom sat on his pallet mending a shirt with thread he'd salvaged from a torn sack. Walsh was writing a letter by the dim light of the single bulb. Miller and Cooper played cards with a deck so worn the suits were barely visible.

Price, the New Zealander, stood beside Dick looking out at the snow. He'd been at the farm since the beginning, a quiet man in his late twenties who rarely spoke unless spoken to. But lately Dick had noticed him watching the fences, studying the guards" routines, his eyes tracking patterns.

"Beautiful, isn't it?" Price said. "The snow."

"Cold," Dick replied.

"That too." Price was quiet for a moment. "Makes you think about home, though. I'm from the South Island. We get snow like this in winter. Real snow, not this northern hemisphere stuff."

Dick smiled slightly. "All snow looks the same to me. Newcastle doesn't get any."

"You're missing out, mate." Price pulled his thin jacket tighter. "How long you reckon this war's going to last?"

The question hung in the cold air. Dick had learned not to think too hard about it. Thinking about the duration led to despair, and despair led to giving up.

"Long as it takes," Dick said finally.

"That's not an answer."

"It's the only answer there is."

Price looked at him, his eyes measuring. Then he turned and went back inside. Dick watched him go, unease working through him. He'd seen that look before, on men who were making plans they shouldn't be making.

The snow continued through the night and into the next day. Work in the fields stopped. The prisoners spent their time in the shed or doing light tasks around the farm buildings. Mending tools, stacking firewood, clearing snow from paths. Werner, the old guard, supervised with his usual quiet efficiency. Beck, the young one, prowled the perimeter looking for infractions that didn't exist.

Dick worked alongside Tom, splitting logs behind the barn. The axe felt good in his hands, the physical work warming him from the inside. Each swing, each crack of wood, was satisfying in a way that went beyond the simple task.

"Price has been talking to Miller," Tom said between swings. "Asking questions."

"What kind of questions?"

"About the fences. About when the guards change shifts. About how far it is to the coast." Tom buried his axe in a log and looked at Dick. "I think he's planning something stupid."

Dick had thought the same thing. "Planning or just thinking?"

"Does it matter? Thinking leads to planning. Planning leads to trying. And trying gets you killed."

"Maybe." Dick picked up another log and set it on the block. "Or maybe it gets you home."

Tom stared at him. "You're not seriously considering it."

"I didn't say that."

"You didn't have to. I can see it on your face." Tom's voice dropped even lower. "Dick, listen to me. Escape from here is impossible. We're in the middle of Poland. Hundreds of miles from anywhere friendly. We don't speak the language. We have no papers, no money, no food. Even if we got past the fences, where would we go?"

"Home," Dick said. "Eventually."

"You'd die in the first week. Freeze to death. Starve. Get caught and shot. This isn't an adventure, mate. This is suicide with extra steps."

Dick knew Tom was right. But there was another part,

the part that pulled out Eileen's letters at night and read them until the words blurred. The part that looked at the photograph of his family and wondered if Meryl even remembered what he looked like anymore. That part whispered that staying here, year after year, was its own kind of death.

"I'm not planning anything," Dick said finally. "I'm just thinking."

"Well stop thinking. Nothing good comes from it."

But Dick couldn't stop. Over the next few days, as the snow continued to fall and work remained suspended, he found himself watching the same things Price was watching. The fence line. The guards" patterns. The distance to the tree line. The gaps in the wire where snow had weighted it down.

Tom was right about the odds. Almost zero. But almost zero wasn't quite zero. And after eighteen months as a prisoner, after eighteen months of hunger and cold and waiting, almost zero started to look like a chance worth taking.

Price approached him on the fourth evening. The others were inside, clustered around the stove, but Dick had stepped out for air despite the cold. Price came and stood beside him, both of them looking at the white fields and the dark tree line beyond.

"I'm going to try," Price said without preamble. "Next week, probably. When the moon's down and the snow's stopped. I wanted you to know."

"Why tell me?"

"Because you've been watching too. Because I think you

understand." Price turned to look at him. "I can't stay here, Dick. I can't spend the rest of the war, however long that is, splitting logs and digging ditches. I'll go mad. I need to try, even if it kills me."

"It probably will kill you."

"I know. But at least I'll have tried. At least I won't die wondering what if."

Dick understood that. The what ifs ate at you in the quiet moments, made you question everything.

"Are you asking me to come with you?" Dick asked.

"I'm telling you my plan. What you do with that information is up to you."

Price laid it out simply. The fence by the north paddock, where the posts had rotted and the wire sagged. The forest beyond, thick enough to provide cover. They'd go at dusk, when the light was failing but before the moon rose. East first, staying off the roads, then south once they were clear of the farm. Find a village, steal food and warmer clothes, keep moving. Make for the coast eventually, find a way onto a ship. It would take weeks, maybe months. But it was possible. Barely possible, but possible.

"The wire by the shed," Price continued. "There's a small hatch for ventilation. I've been working the hinges loose. It opens now, if you're careful. That's how we get out of the shed after lockup."

"We?" Dick asked.

"I'm hoping it's us. But if it's just me, I understand."

Dick thought about Eileen. About his promise to come home. About whether dying in the snow trying to escape was better or worse than slowly dying in captivity. He thought about his mother's letters, about Lorna's sharp tongue, about Meryl asking when he was coming home.

"I need to think about it," Dick said.

"Don't think too long. I'm going next week, with or without you."

That night, Dick lay on his pallet unable to sleep. Above him, Tom's breathing was slow and steady. Around him, the other prisoners snored and shifted. Outside, snow continued to fall.

He pulled out the photograph and looked at it by the moonlight coming through the high window. His family, frozen in time. They were waiting for him. But would they rather he stayed alive as a prisoner or died trying to get home?

He tried to think it through rationally. The odds were terrible. Price was right about that. They might make it to the forest. They might even make it a few miles beyond. But then what? Winter in Poland with no food, no shelter, no papers. German patrols. Polish civilians who might turn them in. The sheer distance to anywhere safe.

And if they were caught, when they were caught, the punishment would be severe. Solitary confinement at minimum. Or they'd be shot during the attempt, like prisoners in other camps he'd heard about.

But staying meant more months, maybe years, of this. Splitting logs and digging ditches and waiting for letters that took months to arrive. Watching himself forget what his sisters looked like. Watching Eileen's face blur in his memory until she was just a name, just words on paper.

Dick put the photograph away. He didn't know what the right choice was. Maybe there wasn't a right choice.

Over the next three days, Dick found himself helping Price without quite admitting he was committed. He tested the ventilation hatch when no one was looking. He studied the fence line during work details. He memorized the guards" patterns, noting when Werner went to the house for meals, when Beck made his rounds, and where the blind spots were.

Tom noticed, of course. Tom always noticed.

"You're really considering this," Tom said. It wasn't a question.

"Maybe."

"Dick, listen to me as your friend. Don't do this. Please. The odds are terrible. You'll die out there."

"Or I'll make it home."

"You won't. No one makes it home from Poland. The geography alone makes it impossible."

"But not impossible."

"Impossible enough." Tom grabbed Dick's arm. "I can't watch you die. I've already watched Jack die. I've watched Mick die. I can't watch you die too."

Dick swallowed. Mick, falling on the hillside during the withdrawal. Jack, shot through the throat on the first day at Maleme. Both gone. Both men who should have made it home but didn't.

"I haven't decided yet," Dick said.

"Yes you have. I can see it in your eyes. You're going to try."

Dick didn't deny it.

The snow stopped on Friday. The sky cleared and the temperature dropped even further. The fields were a perfect white under a hard blue sky. The guards" breath came out in clouds. The prisoners worked with stiff fingers and numb feet, clearing snow from paths and hauling firewood.

Price approached Dick during the midday meal. "Tonight," he said. "Moon sets at eight. Guards change at nine. We go at eight thirty."

Dick went still. "Tonight?"

"The weather's clear. The ground's frozen. It's now or never."

"I need more time."

"There is no more time. I'm going tonight. Are you with me or not?"

Dick looked across the yard at Tom, who was watching them with worried eyes. He looked at Walsh and Miller and Cooper, men who'd become friends through shared misery. He looked at Werner, who'd treated them decently within the confines of his duty. He looked at the fence and the forest beyond.

Then he thought about Eileen, about her last letter. Please come home. That's what she'd written. Not stay safe. Not wait for the war to end. Come home.

He thought about sitting in this shed for another year, another two years, watching the seasons change while his life passed him by. He thought about the photograph in his pocket, about the faces that were already starting to blur.

He thought about dying here, of cold or disease or despair, never having tried.

"I'm with you," Dick said.

Price nodded once. "Pack light. Wear everything you have. Bring nothing that rattles or shines. Be ready at eight."

The afternoon crawled by. Dick worked mechanically, his mind elsewhere. Tom wouldn't speak to him, just worked in angry silence. Dick wanted to explain, to make Tom understand, but there were no words that would help.

At evening lockup, Beck secured the door and walked away. Werner, who usually stood guard through the evening, had gone to the farmhouse for his meal. Through the window, Dick could see lights in the house, could smell food cooking. His stomach growled.

Tom sat beside him on the pallet. "Last chance to change your mind."

"I know."

"They'll shoot you, Dick. They will actually shoot you. That's not a threat or a guess. That's what happens to prisoners who try to escape."

"I know."

"And you're doing it anyway."

"I have to try." Dick looked at Tom. "I know it's stupid. I know the odds. But if I don't try, I'll spend the rest of my life wondering if I could have made it. I can't live with that."

Tom was quiet for a long moment. Then he pulled something from his pocket and pressed it into Dick's hand. A small knife, carved from a piece of metal, with an edge honed against stone. "Take this. If you're going to do something stupid, at least have a weapon."

Dick closed his fingers around the knife. "Tom..."

"Don't thank me. Just don't die. If you die, I'll be furious with you."

"I'll do my best."

The others had noticed something was happening. Walsh and Miller sat nearby, watching. Cooper had stopped writing his letter.

"You're going, aren't you?" Walsh asked.

Dick nodded.

"Christ." Walsh shook his head. "You know it's suicide."

"So I've been told."

"But you're doing it anyway."

"Yeah."

Walsh stood and pulled off his jacket. It was warmer than Dick's, with fewer holes. "Take this. You'll need it."

"Walsh, I can't..."

"Take it. If you freeze to death because I didn't give you my jacket, I'll never forgive myself." Walsh pressed it into Dick's hands. "Now go before I change my mind and tell the guards."

Miller contributed a pair of gloves. Cooper gave him a wool cap. In moments, Dick had more warm clothing than he'd had in months, donated by men who had almost nothing themselves.

"Why?" Dick asked.

"Because you're trying," Miller said. "Most of us have given up. We're just waiting for the war to end, however long that takes. But you're actually trying to get home. That's worth something. That's worth a jacket and some gloves."

Tom looked away, his jaw tight.

At eight o'clock, Price moved to the ventilation hatch. He pushed it open, the hinges silent from weeks of careful work. Cold air rushed in. Price looked at Dick.

"Ready?"

Dick pulled out the photograph one last time. His family. He looked at each face, memorizing them. His mother. His father. Jean, Lorna, Dorrie, Shirley, Meryl. If he died tonight, this would be the last time he saw them.

He put the photograph back in his pocket, right next to Eileen's letters. Then he nodded to Price. "Ready."

Price went through the hatch first, dropping into the snow outside. Dick followed, squeezing through the narrow opening. The cold slapped him. His breath came in white clouds. The snow crunched under his boots.

They moved quickly but carefully, keeping low, heading for the fence line. The moon had set. The only light came from the farmhouse and the distant stars.

The fence was fifty yards away. They covered the distance in a crouch, expecting shouts at any moment. But none came. They reached the wire and Dick saw what Price had meant. The posts were rotten, the wire sagging under the weight of snow. There was almost a foot of clearance at the bottom.

Price dropped flat and started working his way under. The wire caught on his jacket. He twisted, pulled free, kept going. Then he was through, on the other side, standing in the open ground between the fence and the forest.

Dick dropped and followed. The snow soaked through his clothes immediately. The wire scraped across his back. For a moment the wire caught him, held him. He twisted, heard fabric tear, and he was through.

They ran for the tree line. The snow was deeper here, slowing them down. Dick's boots sank with each step. His breath came in gasps. Behind them, the farm sat quiet in the darkness. No shouts. No whistles. They'd made it.

They reached the trees and stopped, both of them breathing hard. Dick looked back at the farm. The shed where Tom and the others waited. The barns and house. Werner and Beck somewhere inside, unaware that two prisoners were gone.

"We did it," Price said, his voice full of disbelief. "We actually did it."

"Not yet. We're not clear yet."

They started moving east, staying in the trees, avoiding open ground. The forest was dense here, the branches heavy with snow. It was dark beneath the canopy, hard to see. Dick stumbled over roots and fallen branches, but he kept moving.

They'd been walking for maybe ten minutes when they heard it. A shout from the farm. Then another. A whistle, sharp and urgent. They'd been discovered.

"Run," Price said.

They ran. Through the trees, crashing through underbrush, branches whipping at their faces. Behind them, more shouts. The sound of men organizing. Dogs barking.

Dogs. Dick hadn't thought about dogs.

The barking grew louder. They changed direction, angling south, trying to put distance between themselves and the pursuit. Dick's lungs burned. His legs were heavy, each step agony. But he kept running because stopping meant capture and capture meant punishment or death.

They broke from the trees into a clearing. Open ground, maybe a hundred yards across. On the other side, more forest. But crossing the clearing meant being exposed. Visible.

Price didn't hesitate. He ran into the open. Dick followed, his boots pounding through the snow. They were halfway across when the shout came from behind.

"Halt! Stehenbleiben!"

They didn't halt. They ran harder. Dick's chest felt like it would burst. Fifty yards to the trees. Forty. Thirty.

The first shot cracked through the night. Dick heard the bullet snap past, somewhere to his right. Price stumbled but kept running. Another shot. Then another.

Twenty yards. Almost there. Dick could see the safety of the trees, the darkness that would hide them.

Price went down.

One moment he was running, the next he was falling, his legs tangling, his body hitting the snow hard. Dick stopped and turned back. Price was trying to get up, his face twisted with pain. Blood spread dark across the snow beneath him.

"Go!" Price shouted. "Keep running!"

Dick hesitated. Behind them, figures were emerging from the trees. The guards. The dogs. They had maybe seconds.

"Go!" Price shouted again. "Don't let this be for nothing!"

Dick ran. He left Price bleeding in the snow. Price was right. If Dick stayed, they'd both be caught. If he ran, at least one of them might make it.

He reached the trees and kept running, branches tearing at him, his breath sobbing in his throat. Behind him, more shots. Then silence. Then a single shot, close and final.

Dick stopped, his back against a tree, his body shaking. That last shot. He knew what it meant. Price was dead. Shot trying to escape, or executed after capture. Either way, dead.

Dick slid down to sit in the snow, his head in his hands. Price was dead. And for what? They'd made it maybe half a mile from the farm. Half a mile before it all went wrong.

He should keep running. Put distance between himself and the guards. But his body wouldn't move. Price was dead. The man who'd planned the escape, who'd convinced Dick to try, who'd believed they could make it home. Dead in the snow in Poland, so far from home.

Time passed. Dick didn't know how long. Long enough for his body to start shaking from cold and shock. Long enough for his mind to start working again.

He had to move. Sitting here meant dying of exposure. He pulled himself up and started walking, keeping to the trees, heading vaguely south. His body moved on autopilot while his mind replayed the escape. The fence. The run through the trees. The clearing. Price falling. That final shot.

Again and again.

He walked for maybe an hour before he heard the dogs again. They'd picked up his trail. Of course they had. The snow made tracking easy. He couldn't outrun them. Couldn't hide from them. It was over.

Dick stopped walking and stood waiting. The barking grew louder. Then flashlights appeared through the trees, bobbing as the men ran. Dick raised his hands, making it clear he wasn't going to fight.

The dogs reached him first, big German shepherds that circled, snarling. Then the guards. Werner first, breathing hard, his face grim. Beck right behind him, his rifle raised, his finger on the trigger.

"Don't shoot," Werner said sharply in German. "We need him alive."

Beck looked like he wanted to shoot anyway. His hands shook on the rifle. His brother's death, Price's death, all of it had pushed him somewhere dark.

Werner moved between Beck and Dick. "Put the rifle down. Now."

Beck lowered it. Werner pulled Dick's arms behind his back and tied them with rope. Done without roughness but without gentleness either. Just efficiency.

"The other one?" Dick asked in German.

Werner's face was stone. "Dead."

Dick had known. But hearing it confirmed was different. Price was dead. Because he'd wanted to go home. Because he'd refused to wait for the war to end. Because Dick had agreed to come with him.

They marched him back through the forest, the dogs leading, the guards following. When they reached the clearing, Dick saw Price's body still lying in the snow. The guards had thrown a blanket over him, but it didn't quite cover everything. One arm extended from beneath it, reaching toward nothing.

Dick looked away.

They took him back to the farm, back through the fence he'd crawled under a lifetime ago. The other prisoners were out in the yard, lined up under guard while the shed was searched. They looked at Dick as he was marched past. Tom's face was stricken. Walsh looked sick. Miller just shook his head.

Werner pushed Dick into a small storage room in one of the barns and locked the door. It was dark and cold, with a dirt floor and walls that smelled of old hay. Dick sat in the corner and tried to stop shaking.

Price was dead. The escape had failed. And Dick was alive only because Werner had stopped Beck from shooting him in the forest.

He pulled out the photograph, his hands numb and

clumsy. Even in the darkness, he knew each face. His family. The people he'd been trying to get home to. The people he'd probably never see again because he'd been stupid enough to try escaping.

Eileen's letters were still in his pocket too, but he couldn't bear to read them. Please come home, she'd written. He'd tried. He'd failed. And Price had died because of it.

Dick closed his eyes and saw it again. Price falling. The blood in the snow. That final shot.

Again and again.

The night stretched on. Dick sat in the darkness, cold seeping into his bones, guilt pressing down on him. Price's face. Price's voice: "Don't let this be for nothing." But it had been for nothing. They'd failed after half a mile.

Dawn came grey and cold. The door opened and Werner stood there with bread and water. He set them on the floor and looked at Dick.

"You will be punished," Werner said in English. "I do not know what form. That is not my decision. But you will be punished."

"What about Price's body?"

"We will bury him. With respect. He was a soldier, even if he was a fool."

"He wasn't a fool. He was brave."

"Bravery and foolishness often look the same." Werner paused at the door. "I am sorry about your friend. I am sorry this happened. But you understand I had no choice. If prisoners escape, I am responsible. My family is responsible. You forced this."

"I know."

Werner left, locking the door again. Dick ate the bread mechanically, tasting nothing. He drank the water. Then he sat and waited for whatever came next.

It came that afternoon. Two military police arrived from the main camp, bringing orders. Dick was to be returned to Stalag VIII-A immediately. His work brigade assignment was terminated. He would face punishment there.

They allowed him to collect his few belongings from the shed. Tom met him at the door, his eyes red.

"I told you," Tom said. "I told you this would happen."

"I know."

"Price is dead because you both wanted to be heroes."

"We weren't trying to be heroes. We were trying to go home."

"Well now Price is dead and you're going to punishment cells. Was it worth it?"

Dick didn't answer because he didn't know. He'd tried. He'd failed. Price had died. Were those facts worth the attempt? He honestly couldn't say.

The other prisoners said quiet goodbyes. Walsh gripped his shoulder. Miller nodded. Cooper looked away. They all knew this could have been any of them. Could still be, if desperation got strong enough.

Werner stood in the yard as the military police loaded Dick into a truck. He met Dick's eyes once. Not quite sympathy. Not quite approval. Just acknowledgment that they'd both done what they had to do, and now they'd both have to live with the consequences.

The truck drove away from the farm. Dick looked back once and saw Tom standing alone in the yard, watching him go. Then the farm disappeared behind hills, and Dick was alone with the military police and the knowledge of what he'd done.

Price was dead. The escape had failed after half a mile. And Dick was going back to the main camp to face punishment he couldn't quite imagine.

But somewhere under the guilt and the grief and the fear, there was one hard truth he couldn't deny: he'd tried. He'd actually tried to escape, to get home, to be more than just a prisoner waiting for the war to end. He'd failed, but he'd tried.

Price had been right about one thing. At least they wouldn't die wondering what if. Price had died trying. And Dick was alive to carry that knowledge forward, to remember, to maybe try again someday if the chance came.

The truck drove on through the grey afternoon. Dick closed his eyes and saw Price one more time, falling in the snow, reaching toward home.

Then he pushed the image away and focused on surviving whatever came next. Because that's all he could do now. Survive. Remember. Wait.

And maybe, someday, try again.

CHAPTER 13 — SOLITARY

The military police delivered Dick to Stalag VIII-A just after dark. The camp looked smaller than he remembered, the barracks grey and identical in the failing light, the guard towers stark against the sky. He'd been gone for four months. It felt like years.

They marched him straight to the commandant's office, a small building near the main gate with electric lights that hurt Dick's eyes after hours in the truck. The commandant was a thin man in his fifties with steel-rimmed glasses and the bearing of someone who'd spent his entire life in the military. He looked at Dick without emotion, just assessment.

"Corporal Roberts," the commandant said in accented but precise English. "You attempted to escape from a work brigade. One man died in the attempt. You were recaptured within two hours. Do you have anything to say?"

Dick stood at attention, his hands at his sides. "No sir."

"No explanation? No excuse?"

"No sir."

The commandant studied him for a long moment. Then he opened a file on his desk and made a note. "Seven days solitary confinement. Reduced rations. No letters, no parcels, no contact with other prisoners. After that, you return to the general population. But your work brigade privileges are permanently revoked. You will remain in this camp for the duration of the war. Do you understand?"

"Yes sir."

"One more thing, Corporal. The man who died, Price. His family will be notified. But they will be told he was killed attempting to escape. They will not be told about your involvement. That is a mercy I am granting you. Do not make me regret it."

"Yes sir."

The commandant nodded to the guards. "Take him to the punishment block."

The punishment block was a separate building behind the main barracks, a long, low structure with small cells built into its length. The guards led Dick to the third cell from the end and unlocked the heavy wooden door. Inside was darkness and the smell of damp earth.

"Inside," one of the guards said.

Dick stepped through the door. It closed behind him. The lock turned. Footsteps retreated. Then silence.

Dick stood in the darkness, letting his eyes adjust. The cell was perhaps six feet by eight feet. A dirt floor. Stone walls. A wooden bucket in the corner. No window except a narrow gap near the ceiling where the wall met the roof. Through that gap came a thin shaft of moonlight, barely enough to see by.

No bunk. No blanket. No light. Nothing except the

bucket and the four walls.

Dick sat down with his back against the wall. The cold from the stone seeped through his clothes immediately. His ribs ached from where Beck had hit him with his rifle butt during the recapture, a blow Dick had forgotten to mention to the commandant. His hands were scraped raw. His feet throbbed with blisters from running through the snow.

Price was dead.

Dick closed his eyes and saw it again. Price running across the clearing. The shots. Price falling. The blood dark on the snow. That final shot after Dick had run into the trees.

He should have stayed. Should have gone back for Price. Should have done something other than run away and leave his friend dying in the snow. The fact that staying would have meant his own death didn't make it better. It just meant he was alive and Price wasn't.

He pulled out the photograph from his inside pocket. Even in the dim light from the gap near the ceiling, he knew every face. His mother. His father. His five sisters. The backyard in Alfred Street. A moment frozen in time, from a world that seemed impossibly distant now.

What would his mother think if she knew what he'd done? If she knew he'd tried to escape, that a man had died because of that decision, that Dick had run away instead of staying to help? Would she understand? Or would she be ashamed?

He thought about Lorna. She'd understand, probably. Lorna always saw through the comfortable lies to the hard truths underneath. She'd know why he'd tried, why he'd had to try, even if it was stupid and doomed from the start.

Dick put the photograph away and pulled out Eileen's letters. Three of them now, folded together, the paper soft from handling. He couldn't read them in the darkness, but he knew the words by heart.

Please come home.

He'd tried. He'd failed. And Price had died.

Dick leaned his head back against the wall and closed his eyes. The cold was already working into his bones. This was going to be a long week.

The first night was the worst. The temperature dropped well below freezing and Dick had no way to stay warm. He curled into a ball on the dirt floor, his arms wrapped around himself, his teeth chattering. Sleep came in short bursts, interrupted by shivers so violent they woke him.

At some point during the night, he heard sounds from the other cells. Coughing from one direction. Someone crying quietly from another. A man shouting in what sounded like Russian or Polish before being told harshly to shut up. Dick wasn't alone in this building, but he might as well have been.

Dawn came grey and dim through the gap in the wall. Dick uncurled slowly, his joints stiff, his body aching. He relieved himself in the bucket and tried not to think about the smell that would accumulate over the next week.

Food came around eight o'clock. The door opened and a guard Dick didn't recognize shoved a dented tin bowl through the gap. Thin soup, mostly water with a few pieces of turnip floating in it. A crust of bread, hard as stone. The door closed before Dick could see who'd brought it.

He ate the bread first, chewing to make it softer, to make it last. His mouth was dry and the bread stuck to his teeth. The soup he drank, forcing himself to take small sips instead of gulping it down. The warmth of it, however minimal, was precious.

When the bowl was empty, Dick set it by the door and tried to stand. His legs wobbled. The cold and the lack of food over the past day had weakened him more than he'd realized. He leaned against the wall until his balance returned.

The day stretched out with nothing to mark its passage except the thin shaft of light that moved across the wall as the sun tracked overhead. Dick tried to keep his mind occupied. He counted steps as he paced the small cell. Three steps one way, turn, three steps back. Over and over. Movement generated heat, however small.

He recited poems he'd learned in school, the words coming back in fragments. He sang songs under his breath, old songs from before the war. Anything to fill the silence, to keep his mind from spiraling into dark places.

But the dark places came anyway. He thought about Price, about the clearing, about the blood in the snow. He thought about Tom's face when Dick had said he was going to try escaping. Tom had known. Tom had tried to warn him. And Dick had gone anyway.

He thought about the other prisoners at the farm, about

Walsh giving him his jacket, about Miller giving him gloves. They'd helped him escape, and now Dick was in solitary while they were still at the farm, probably facing questions about what they knew and when they knew it.

He thought about Werner, about the old guard's face when he'd found Dick in the forest. There had been no anger there, just weariness. Just a man doing a job he didn't want to do, in a war he didn't want to fight.

And he thought about Beck, about the younger guard's hands shaking on his rifle, about how close Dick had come to being shot in that forest. If Werner hadn't stopped him, Dick would be dead beside Price. Two bodies instead of one. Two families receiving telegrams. Two more names added to the endless list of the dead.

The afternoon brought more soup and another crust of bread. The same guard, the same routine. Door opens, the bowl shoved through, the door closes. No words, no acknowledgment that Dick was human.

Dick ate mechanically, his body going through the motions while his mind was elsewhere. He was starting to understand what solitary confinement actually meant. It wasn't just being alone. It was the absence of everything that made you feel human. No conversation. No touch. No variation in routine. Just the cell and the cold and the endless silence.

By the second night, Dick's thoughts had started to blur. The cold never left. He couldn't remember what it felt like to be warm. His hands were numb. His feet were numb. Only his core still generated any heat, and even that was fading.

He dreamed about Newcastle, about walking on the beach with the sun hot on his shoulders. He dreamed about his mother's kitchen, about the smell of food cooking, about sitting at the table with his sisters arguing around him. He dreamed about Eileen, about her face, about her voice on the telephone exchange, calm and precise.

When he woke, the cell was dark and he couldn't remember where he was for a moment. Then reality returned, hard and cold.

He forced himself to stand, to pace, to count steps. Movement. Focus. Don't let the cold and the isolation win. He'd survived Maleme. He'd survived the withdrawal. He'd survived the journey through Warsaw. He could survive seven days in a cell.

But Price hadn't survived. That thought kept returning, no matter how hard Dick tried to push it away. Price had wanted to go home. He'd been willing to risk everything for the chance. And it had killed him.

Was Dick any different? He'd made the same choice, taken the same risk. The only difference was luck. Dick had been slower, had fallen behind, had reached the trees while Price was caught in the open. If their positions had been reversed, it would be Dick's body buried somewhere near that farm, and Price sitting in this cell.

The randomness of it was suffocating. Life and death decided by seconds, by small choices that seemed meaningless at the time. Run left or run right. Stay or go. Live or die. There was no logic to it, no fairness. Just chaos.

The third day brought a visitor. The door opened and Werner stood there, his rifle slung over his shoulder, a tin bowl in his hands. He looked at Dick for a long moment before setting the bowl down and stepping inside the cell.

"You," Werner said in English. "You make trouble for everyone."

Dick said nothing. Speaking took energy he didn't have.

"The commandant, he wants to shoot you," Werner continued. "But I speak for you. I tell him you are not in your right mind. I tell him solitary confinement has broken something in your head." Werner pulled something from his pocket and held it out. A piece of bread, larger than the usual ration. "I do not know if this helps. But I try."

Dick took the bread with numb fingers. "Thank you."

"Do not thank me. Just survive this. Do not make me speak for you again." Werner paused at the door. "My son, he is your age. Somewhere in Russia now. I do not know if he is alive. When I look at you, I think about him. I think maybe someone shows him mercy also. So I show mercy to you. But do not waste it."

After Werner left, Dick stared at the bread, this small act of kindness from a man whose son might be dead in Russian snow. Werner had lost his brother in the last war. Might lose his son in this one. And still he showed mercy.

Dick ate the extra bread, making it last. Werner had shown him mercy.

The fourth day blurred into the fifth. Dick lost track

of which was which. His body had stopped generating heat properly. He found himself staring at the wall. How long had he been staring? He couldn't remember when he'd started.

He pulled out the photograph again and looked at it. But the faces seemed strange now, distant, like people from a story rather than his actual family. How long since he'd seen them? A year? Eighteen months? Time had lost its meaning.

He tried to remember Eileen's face, the exact shape of it, the color of her eyes. But the memory wouldn't come clear. He knew she existed. He knew he loved her. But the specifics were fading, worn away by cold and isolation and time.

On what he thought was the fifth night, Dick had a dream that felt more real than reality. He was back in Newcastle, walking down Alfred Street toward the beach. His mother was in the front garden, hanging washing. She looked up and smiled at him.

"You're home," she said.

"Yes," Dick replied.

"It's good you're home. We've been waiting."

He walked past her to the front door. Inside, his sisters were setting the table for dinner. Jean and Lorna and Dorrie and Shirley and Meryl, all of them talking at once, their voices overlapping. His father sat in his chair by the window, reading the newspaper, ink on his fingers.

"Where have you been?" Lorna asked.

"Away," Dick said.

"Well you're back now. Sit down. Dinner's ready."

He sat at the table and they passed food around. Roast lamb and potatoes and vegetables. His mother served everyone with her steady hands. The food was hot and delicious and Dick ate until he was full.

Then Eileen was there, though Dick hadn't seen her come in. She stood by the window, looking out at something he couldn't see.

"Are you staying?" she asked without turning around.

"I don't know," Dick said.

"You should decide. You can't be in two places at once."

Dick woke with a start, his body shaking with cold. The cell was dark. He was alone. The dream faded, leaving only a residue of warmth that made the reality worse by comparison.

Dick forced himself to stand, to move, to prove he was still here, still real, still himself. But the effort was enormous. His legs barely held him. His head swam. He sat back down before he fell.

The days continued. Dick counted meals. Three per day, except when they forgot or decided he didn't need one. He tried to calculate how long he'd been here, but the numbers kept slipping away.

On the sixth day, Dick stopped eating. Not as a protest. His body had simply given up. The soup came and sat by the door, untouched. The bread lay beside it. Dick looked at them but couldn't summon the energy to reach for them.

He lay on the dirt floor and stared at the gap near the ceiling where the light came through. He thought about

Price, about whether Price had felt this same emptiness in those final moments. Whether he'd regretted trying. Whether he'd thought about home as the life left him.

Dick thought about Tom, about the last conversation they'd had at the farm. Tom had been right. The escape had been suicide. Price was dead and Dick was dying by degrees in this cell. Nothing had been gained. Nothing had been worth it.

Except he'd tried. He'd actually tried. That had to mean something. Even if it had failed, even if it had cost Price his life, at least they'd tried. At least they hadn't just sat and waited for the war to end.

But was that enough? Was I trying enough to justify the cost?

Dick didn't know.

The door opened. Werner stood there, looking down at Dick on the floor. His face shifted from stern to concerned.

"You must eat," Werner said. He picked up the bowl of soup and knelt beside Dick with some difficulty. "Come. Eat. Or you die."

"Maybe that's better," Dick said.

"No. No, it is not better." Werner lifted Dick's head and tipped the bowl against his lips. "Drink. Please."

The soup ran into Dick's mouth. He swallowed reflexively. Werner kept tipping the bowl until it was empty, then picked up the bread and pushed it into Dick's hands.

"Eat. You must eat. I spoke for you to commandant. I tell him you will survive. Do not make me liar." Werner's voice dropped. "My son. I think about my son. I cannot watch you die. Please."

Dick bit into the bread. It was like eating sawdust, but he chewed and swallowed. Werner watched until the bread was gone, then nodded and stood.

"Tomorrow you go back. Tomorrow is done. Just hold on one more day."

After Werner left, Dick lay on the floor. Werner had saved him.

The world made no sense anymore. Enemies showed mercy. Friends died for no reason.

The seventh day came. Dick marked it by the light, by the meals, by the fact that Werner had said tomorrow. This was the last day. Tomorrow he'd be released back into the general population. Tomorrow he'd see Tom again, and Fletcher, and the others. Tomorrow he'd rejoin the world of the living.

If he survived tonight.

The final night was the longest. Dick lay on the floor, too weak to pace anymore, too cold to sleep. He counted his breaths. Counted his heartbeats. Counted anything countable to prove time was still passing, that dawn would come, that this would end.

He thought about all the people he'd lost. Jack, shot through the throat at Maleme. Mick, bleeding out on a hillside during the withdrawal. Price, falling in the snow. All of them are gone. All of them dead while Dick survived.

Why? What made Dick special? What made him worthy of survival when they weren't? He couldn't answer that question. Couldn't find any logic in it. He was alive because of luck and circumstance and timing. Nothing more. Nothing meaningful.

But he was alive. That was the simple fact. He'd survived Maleme and the withdrawal and Warsaw and the farm and the escape attempt and seven days in solitary. He was still here, still breathing, still himself. Damaged, maybe. Changed, certainly. But alive.

Dawn came grey through the gap in the wall. Dick watched the light strengthen, watched it chase the shadows from the corners of the cell. He'd made it. Seven days. The punishment was over.

The door opened and guards stood there, different ones this time. "Roberts. You're done. Come out."

Dick tried to stand. His legs wouldn't support him. The guards came in and pulled him up, half carrying him out of the cell into the shocking brightness of morning.

The light stabbed his eyes. He turned his face away, squinting, his eyes watering. The air outside the cell felt vast, infinite, too much space after the tiny room. Sound rushed at him: boots on gravel, voices calling, a door slamming somewhere. After seven days of near-silence, it was overwhelming.

They took him to the washhouse first. Let him clean himself with cold water and a sliver of soap. The water on his skin felt strange, almost painful. His reflection in a scrap of mirror showed a face he barely recognized. Gaunt. Hollow-eyed. Grey-skinned. He looked like he'd aged ten years in seven days.

Then they took him to the barracks. Fletcher met him at the door, his face creased with concern. "Christ, Roberts.

You look terrible."

"Feel worse."

Fletcher guided him to his old bunk. Someone had kept it for him, and hadn't let anyone else claim it. Tom appeared from somewhere, his face a mixture of relief and anger.

"You stupid bastard," Tom said. "You nearly died in there."

"I know."

"Was it worth it? Price dead and you half dead. Was it worth it?"

Dick lay down on the bunk, his body finally able to rest. "I don't know. Ask me in ten years."

Tom shook his head but sat down beside him. "We thought you were going to die. Werner told one of the other guards you'd stopped eating. We thought you'd given up."

"I nearly did."

"What stopped you?"

"Werner. He made me eat. Forced soup down my throat." Dick looked at Tom. "The guard whose son might be dead in Russia force-fed me to keep me alive. That's the world we live in now."

Tom didn't have an answer for that. He just sat there while Dick closed his eyes and let exhaustion pull him under. This time, at least, he slept without dreams. Just darkness and the knowledge that he'd survived. That was enough. For now, that was enough.

When he woke hours later, the barracks was quiet. Most

of the men were at work detail. Only the sick and the injured remained. Dick sat up, his body protesting every movement.

Fletcher brought him soup, real soup this time, with actual pieces of meat and vegetables. "Eat slowly. Your stomach won't be able to handle much."

Dick ate, forcing himself to take small bites, to not rush. The food was the most delicious thing he'd ever tasted, though objectively it was still prison camp rations.

"What happened out there?" Fletcher asked. "At the farm?"

"We tried to escape. Price and I. We made it about half a mile before they caught us. Price was shot. I surrendered."

"Why'd you try? You had to know it was hopeless."

"Because staying felt like dying slowly. Escaping felt like dying with purpose." Dick set down the empty bowl. "I was wrong, probably. Price is dead. I nearly died. Nothing was gained."

"Except you tried," Fletcher said quietly. "Most of us just accept this. We sit and wait and hope the war ends before we die of boredom or disease or despair. But you actually tried to get home. That's something. That's worth something."

Dick pulled out the photograph, checking that it had survived the week in solitary. It had, though the paper was more worn than ever. His family looked back at him, their faces a reminder of what he was holding onto, what he was surviving for.

"My mother would tell me I was stupid," Dick said. "My

father would tell me to do my duty and stop causing trouble. But Lorna... Lorna would understand. She'd know why I had to try."

"You got any brothers or sisters, Fletcher?"

"One sister. Back in Manchester. Haven't heard from her in six months. Don't know if the letters aren't getting through or if she's stopped writing." Fletcher stood. "Get some rest. You'll need your strength. And Roberts? Don't try that again. Once is brave. Twice is stupid."

Dick nodded. He had no intention of trying again. Not soon, anyway. The solitary confinement had broken something in him, some part that believed escape was possible. What remained was just the simple determination to survive, one day at a time, until the war ended or he died.

That night, lying in his bunk with Tom snoring above him, Dick thought about Price one more time. About the clearing and the blood and that final shot. He thought about Werner forcing soup down his throat. He thought about the commandant's cold assessment.

He thought about Eileen waiting in Newcastle, about whether she'd still recognize him when he finally made it home. If he made it home. The if was bigger now, more real, after seven days in solitary.

But he was still alive. Still breathing. Still himself, more or less. And that would have to be enough. Because that was all he had. That and the photograph and the letters and the thin thread of hope that connected him to the life he'd left behind.

The war continued. The camp continued. And Dick continued with them, one day at a time, carrying the weight of what he'd done and what it had cost. Price was dead. But Dick was alive. And somehow, he'd have to find a way to live with that.

CHAPTER 14 — THE GROUP ESCAPE

Dick returned to work detail on a grey morning in late January. Two months had passed since his release from solitary, two months of his body recovering while his thoughts spiraled darker. He stood in the parade ground for roll call, his breath steaming in the cold air, his shoulder still aching where Beck had hit him with the rifle butt.

Tom stood beside him, close enough that their shoulders nearly touched. Neither spoke. They'd already said everything that mattered.

Hartmann called out work assignments. Dick's name was called for drainage work. He fell in with the group being marched out through the gate. Tom was assigned to coal detail. They made eye contact once before the groups separated.

Tom's expression was clear: Don't do anything stupid.

Dick looked away.

The work was miserable. Clearing frozen ditches, the mud like iron, each strike of the shovel jarring his arms. His wounded shoulder screamed. His hands, weakened from solitary, could barely grip the handle. But he kept working because stopping meant punishment, and he'd had enough punishment to last a lifetime.

Except it wasn't enough. Nothing was enough. Because every night Dick lay in his bunk and thought about Price falling in the snow. Every night he pulled out the photograph and looked at faces that were starting to seem less real, less connected to him. Every night he knew that staying here meant dying slowly.

He was going to try again. He'd known it even as he'd promised himself in solitary that he'd accept his situation. The promise had been a lie the moment he'd made it.

But this time would be different. This time he wouldn't go alone.

Walsh approached him three nights later. The barracks was quiet, most men already asleep. Walsh sat on the edge of Dick's bunk and spoke in a whisper.

"You're planning something," Walsh said. "I can see it on your face."

Dick considered lying. Then he nodded.

"Another escape attempt."

"Yes."

Walsh was quiet for a moment. "They nearly killed you last time. Price did die last time."

"I know."

"And you're going to try again anyway."

"I have to."

Walsh looked at him, studying his face in the dim light from the single bulb. Then he said, "Take me with you."

Dick went still. "What?"

"Take me with you. If you're going to try, I want to come."

"Walsh, you saw what happened to Price. You saw what happened to me."

"I did. I also saw you try when everyone else just sits and waits. I also saw you actually believe you could make it home." Walsh paused. "I can't sit here anymore, Dick. I can't wait for this war to end. My father died three months ago. The letter said he had a heart attack. I wasn't there. I'll never see him again. And every day I sit here is another day my mother's alone."

Dick understood that. Walsh's father was dead and Walsh hadn't been there. The guilt of it, the knowledge that life continued without you, that people suffered and died while you sat trapped.

"Two people are harder to hide than one," Dick said.

"Two people can help each other. Can keep watch while the other sleeps. Can share food." Walsh leaned closer. "You need help, Dick. Last time you went with Price and it nearly worked. Going alone is suicide."

"Going with someone is risking their life."

"My life is at risk." Walsh's voice was firm. "So either you take me with you, or I try on my own. Your choice."

Dick thought about it. Walsh was right that two had advantages. And Walsh was practical, resourceful, could fix things and figure things out when plans went wrong. If Dick was going to try again, Walsh would be good to have along.

"All right," Dick said finally. "But we do it my way. We

plan it properly. We don't rush."

"Agreed." Walsh held out his hand. Dick shook it.

The next night, Walsh brought Miller. Miller sat on Tom's empty bunk, his face uncertain.

"Walsh says you're planning an escape," Miller said.

"Maybe."

"He says you'll take people with you."

Dick looked at Walsh, who shrugged. "Miller's got skills we need. He worked for a surveying company before the war. He can read maps, navigate by stars. He knows distances and terrain."

Miller nodded. "I do. And I want to come."

"Why?" Dick asked.

"Because I'm twenty-four years old and I've spent a year and a half in this place. Because my girl probably thinks I'm dead. Because if I don't try, I'll spend the rest of my life wondering if I could have made it." Miller paused. "Same as you, I reckon."

Dick looked at both men. Walsh, practical and determined. Miller, young and desperate. Both willing to risk everything for a chance at home.

"Three people are even harder to hide than two," Dick said.

"Three people can watch in shifts," Walsh replied. "Three people can carry more supplies. Three people have better odds if one gets hurt."

Or three people die instead of one, Dick thought. But he didn't say it.

"If we do this," Dick said, "we do it right. We plan every detail. We study the routes, the timing, the guards. We don't go until we're ready."

"Agreed," Walsh said. Miller nodded.

"And if we get caught, you don't tell them it was my idea. You say you convinced me. Understand? They already think I'm unstable. Let them think you corrupted me, not the other way around."

"Dick..." Walsh started.

"That's the condition. Take it or I'll go alone."

Walsh and Miller exchanged looks. Then Walsh nodded. "All right. We agree."

Tom discovered the plan a week later. Dick had been studying the duty rosters that Fletcher sometimes left lying around, trying to memorize guard patterns. Tom saw him and put it together immediately.

Tom pulled him aside after the evening roll call, his face tight with anger.

"You promised," Tom said.

"I know."

"You nearly died last time. Price did die."

"I know."

"And now you're taking Walsh and Miller with you." Tom's voice shook. "You're going to get them killed too."

"Or we'll make it."

"You won't. The odds are terrible. You know the odds." Tom grabbed Dick's arm. "I've already watched Jack die and Mick die and Price die. I can't watch you die. I can't watch Walsh and Miller die. Please, Dick. Please don't do this."

Dick pulled his arm free. "I have to try."

"Why? Why do you have to throw your life away?"

Dick pulled out the photograph. The paper was so worn now it was translucent in places, threatening to tear along the creases. "Because they're waiting for me. Because every day I'm here is a day I'm not there. Because if I don't try, I'll die here anyway, just slower."

Tom looked at the photograph, at Dick's family frozen in time. Then he looked at Dick, and his expression shifted from anger to something sadder, something like resignation.

"When?" Tom asked quietly.

"Early March. When the weather improves but before the spring work gets heavy. The guards will be spread thin."

Tom was silent for a long moment. Then he reached into his pocket and pulled out a folded paper. "Miller drew this. Route to the Swiss border. It's not perfect, but it's better than nothing."

Dick took the map with shaking hands. "Tom..."

"Don't thank me. Just make it. If you're going to do this insane thing, then make it all the way. Don't die for nothing." Tom's voice broke slightly. "And if you make it, you tell my family I'm alive. You tell them I'll come home when the war ends. You tell them I'm all right."

"I will."

"Swear it."

"I swear."

Tom nodded once and walked away, his shoulders hunched. Dick watched him go and understood what that map had cost. Tom didn't believe they'd make it. But he was giving them the best chance he could anyway, because that's what friends did.

They planned for three weeks. Dick, Walsh, and Miller met in quiet corners, speaking in whispers, studying the map Tom had given them. Miller traced routes with his finger, calculating distances, noting landmarks.

"Eighty miles to the Swiss border," Miller said. "If we can do twenty miles a day, four days. But we won't be able to do twenty miles a day. More like ten, maybe fifteen if we're lucky. So figure a week, maybe ten days."

"Food?" Walsh asked.

"We'll have to forage. Steal from farms. Eat raw vegetables if we have to." Dick had done this before, and knew what it took. "Water's easier. Plenty of streams this time of year."

"Guards?" Miller asked.

"Patrols on the main roads. Dogs if they track us from camp. German soldiers near the border." Dick had thought about this constantly. "We stay off roads. We move at night when possible. We avoid villages and farms unless we're desperate."

"When do we go?" Walsh asked.

Dick looked at the calendar Fletcher kept on the wall. Early March. The snow was melting. The days were getting longer. Soon the spring planting would begin and work details would be everywhere.

"Next week," Dick said. "Tuesday. There's a supply delivery scheduled. The guards will be distracted."

Walsh and Miller nodded. It was decided.

The night before the escape, Dick lay in his bunk and tried to sleep. Above him, Tom was silent. Dick knew Tom was awake too, knew he was thinking about tomorrow, about what might happen.

"Tom?" Dick said quietly.

"Yeah?"

"Thank you. For the map. For everything."

Tom didn't answer for a long moment. Then he said, "Just make it, Dick. Make all of this worth something."

"I'll try."

"No. Do it. Make it all the way home. Or don't go at all."

Dick closed his eyes and tried to believe that was possible.

Tuesday came grey and cold. Dick worked through morning roll call and breakfast, his body moving automatically while his mind counted down the hours. The supply wagon was due at midday. That's when they'd go.

Walsh and Miller were on the same work detail, clearing timber from the forest edge. The guards were two older men, reservists who moved slowly and watched half-heartedly. When the supply wagon arrived, both guards went to check the manifest.

Dick caught Walsh's eye. Nodded once.

They walked away from the timber pile, moving with purpose but not urgency. Dick's heart hammered so hard he thought the guards would hear it. They reached the tree line and slipped into the shadows.

No shouts. No whistles. They kept walking, picking up speed gradually.

They were a hundred yards into the forest when Miller whispered, "I think we made it."

"Not yet," Dick said. "Keep moving."

They walked until Dick's legs trembled and his lungs burned. They walked until the camp was miles behind and the forest was thick around them. Then Dick called a halt and they collapsed against trees, gasping.

"We did it," Walsh said, his face split with a grin. "We actually did it."

"First part," Dick said. "Now comes the hard part."

They pulled out the map and studied it by failing afternoon light. Miller traced their route with his finger.

"We're here, roughly. We need to go southwest, following this river valley. Stay off roads. Travel by night when we can." Miller looked up. "First goal is to get twenty miles away before they realize we're gone."

"When's roll call?" Walsh asked.

"Evening. Maybe five hours." Dick stood. "Let's move."

They walked until dark, then kept walking by moonlight. Miller navigated by stars, keeping them on course. Dick's shoulder ached. His feet blistered. But he kept moving because stopping meant capture.

Around midnight they stopped to rest in a dense thicket. They had no food, no water except what they'd found in a stream. They were cold and exhausted and

Dick's body wanted to give up.

But they'd made it through the first day. That was something.

"You think they're looking for us yet?" Miller asked.

"Oh yes," Dick said. "Dogs, patrols, the lot. They'll be furious. Three prisoners escaping at once."

"Will they catch us?" Walsh asked.

Dick thought about Price, about the clearing, about that final shot. "If we're unlucky."

"Then we'd better be lucky," Walsh said.

They slept in shifts, one man always awake and watching. When dawn came grey and cold, they started walking again.

The second day was harder. They were hungry now, properly hungry, their bodies burning through reserves they didn't have. They found a farm field with winter cabbage and pulled up three heads, eating them raw, the leaves tough and bitter but food.

Water they got from streams. Dick drank until his stomach hurt, knowing they might not find water again for hours.

They saw their first German patrol that afternoon. Four soldiers walking along a road, rifles slung casually. Dick and the others pressed themselves into a ditch and didn't breathe until the patrol passed.

"Too close," Miller whispered.

"We're fine," Dick said. But his hands were shaking.

The third day brought rain, cold and steady. They were soaked within an hour, their clothes heavy, their boots squelching with each step. They couldn't light a fire for warmth, couldn't stop moving. Dick's teeth chattered so hard he thought they'd break.

But the rain was also cover. Harder for dogs to track them. Harder for patrols to see them. They kept moving through the downpour, miserable but grateful.

That night they found a barn and crept inside. The barn was empty except for hay and the smell of animals. They burrowed into the hay and slept, three bodies pressed together for warmth.

Dick woke to grey dawn light and realized this was the longest he'd made it in an escape attempt. Three days. Longer than with Price. They were still free. Still alive. Maybe they actually had a chance.

The fourth day, Dick woke feeling almost hopeful. They'd covered maybe forty miles. Halfway to the border. Walsh was in good spirits, making quiet jokes. Miller was confident with the map, sure of their route. They were going to make it.

Then Miller twisted his ankle.

They were crossing a stream, stepping from rock to rock, when Miller's foot slipped. He went down hard, his ankle turning beneath him. The crack was audible even over the rushing water.

Miller gasped and went white. Dick and Walsh pulled him to the bank. His ankle was already swelling, the joint at an ugly angle.

"Is it broken?" Walsh asked.

Miller prodded it carefully, wincing. "Sprained. Badly. Maybe fractured. I can't tell."

"Can you walk?" Dick asked.

Miller tried to stand. He managed two steps before his leg gave out and he collapsed. "No. I can't put weight on it."

Dick and Walsh looked at each other. They needed to move fast, needed to cover ground. A man who couldn't walk meant they were trapped.

"We rest here today," Dick said. "Let the swelling go down. Try again tomorrow."

They found a hollow beneath a fallen tree and made camp. Dick fashioned a crude splint from sticks and strips torn from his shirt. Miller's face was grey with pain but he didn't complain.

They waited. The day crawled by. Miller's ankle didn't improve. By nightfall it was purple and swollen to twice its normal size.

"We have to keep moving," Walsh said quietly to Dick, away from Miller. "We can't stay here."

"I know."

"He can't walk."

"I know."

Walsh looked at him. "So what do we do?"

Dick thought about Price. About leaving him bleeding in the snow. About that final shot. He thought about the promise he'd made to himself: no more deaths on his conscience.

"We help him," Dick said. "We carry him if we have to."

"Dick, that's suicide. We'll be caught for sure."

"Then we'll be caught. But I'm not leaving him."

Walsh stared at him for a long moment. Then he nodded. "All right. We stick together."

On the fifth day, they tried to move. Walsh and Dick took turns supporting Miller, his arms over their shoulders, hopping on his good leg. They managed maybe two miles before all three were exhausted.

They were too slow. Much too slow. At this rate they'd need three more weeks to reach the border. They didn't have three more weeks. They didn't have three more days.

On the sixth day, Miller developed a fever. His skin was hot to the touch, his eyes glassy. The ankle was infected, Dick realized. The swelling had spread halfway up his calf.

"Leave me," Miller said. His voice was weak. "You two go on. I'll be all right."

"No," Dick said. "We don't leave people behind."

"Dick..."

"No."

But Dick knew they were trapped. Miller couldn't walk. The fever was getting worse. They were out of food. Patrols were searching for them. It was only a matter of time.

The seventh day, the patrol found them.

Dick heard the dogs first, barking in the distance. Then voices, German voices, calling to each other. He looked at Walsh. Walsh's eyes were wide, his face pale.

"Run," Miller said. "Both of you. Leave me. Run."

Dick shook his head. "No. We're in this together."

The dogs came first, big German shepherds that circled the hollow, snarling. Then the soldiers, six of them, rifles raised.

"Raus! Come out!"

Dick helped Miller stand, supporting his weight. Walsh came to Miller's other side. Together, the three of them emerged from the hollow, hands raised.

The German sergeant looked at them with disgust. "Three escaped prisoners. Two weeks you have been gone. You have caused much trouble."

"We surrender," Dick said in German.

"Yes. You surrender." The sergeant gestured with his rifle. "Walk. Back to camp."

They walked. Miller between Dick and Walsh, half-carried, his face twisted with pain. The dogs followed, the guards followed. Dick looked back once at the forest they'd come through, at the seventy miles they'd covered. So close. They'd been so close.

The march back took two days. Miller's fever worsened. On the second morning, he could barely stand. The guards put him in a wagon. Dick and Walsh walked behind, exhausted, defeated.

They reached the camp at dawn. The gates opened and they were marched through to the parade ground. The entire camp had been assembled. Hundreds of prisoners standing in formation, watching.

The commandant stood on a platform, his face like stone. Dick, Walsh, and Miller were brought before him.

"Three prisoners attempted to escape," the commandant announced in English. "They were caught after two weeks. This is what happens to those who try to escape."

He looked at each of them in turn. "Private Miller. One week solitary confinement. Private Walsh. One week of solitary confinement." Then his eyes locked on Dick. "Corporal Roberts. Two weeks solitary confinement. This is your second escape attempt. The punishment is clear. If you try again, you will be shot on sight. Do you understand?"

"Yes sir," Dick said.

"I hope so, Corporal. Because the next time we have this conversation will be the last."

They were taken to the punishment block. Dick watched as Walsh and Miller were led to their cells. Miller could barely walk. Walsh looked back once, his face pale but determined.

Then Dick was pushed into his own cell. The door closed. The lock turned. Darkness.

Dick sat with his back against the wall and tried to process what had happened. Two weeks. They'd lasted two weeks. Seventy miles. Almost to the border. Almost free.

But almost wasn't good enough. Almost meant failure. Almost meant punishment.

Almost meant Dick had led two more men into suffering.

He pulled out the photograph with shaking hands. The paper was barely holding together now, the creases worn through, the blood stains from his old wound faded to brown. His family looked back at him.

Price had died because Dick convinced him to try. Walsh and Miller were in solitary because Dick convinced them to try. How many more people would suffer because of Dick's refusal to accept his situation?

The answer came to him in the darkness of the cell, clear and final: No more.

If Dick tried again, and he knew he would, he would go alone. No more partners. No more groups. Just him. If he died, only he died. If he was caught, only he was punished.

He couldn't carry any more guilt. Couldn't watch any more friends suffer for his choices.

Next time, if there was a next time, he'd go alone.

The days blurred together. Cold and darkness and the thin soup that came twice a day. Dick marked time by counting meals. Fourteen days. Twenty-eight meals. He could survive this. He'd survived worse.

On the seventh day, he heard voices from other cells. Walsh and Miller being released. Their week was done. They'd go back to the barracks, back to work, back to the endless waiting.

Dick had another week to go. Another week alone with his thoughts and his guilt.

He thought about Tom, about the look on Tom's face when he'd given them the map. Tom had known they wouldn't make it. Had known and helped anyway. Dick owed him. Owed him to survive, to get home, to tell his family Tom was alive.

He thought about Fletcher, about all the men in the camp who just accepted their situation and waited. Maybe they were wiser than Dick. Maybe acceptance was survival and resistance was just slow suicide.

He thought about Eileen, about whether she was still waiting. Three months without a letter from her now. Maybe she'd given up. Maybe she'd found someone else. Maybe Dick was holding onto a dream that had already died.

But even thinking that, even acknowledging that possibility, didn't change what he knew he'd do. He'd try again. One more time. Alone. Because staying here until the war ended was impossible. His mind wouldn't survive it.

On the fourteenth day, the door opened. Light stabbed his eyes. Guards pulled him out, half-carried him to the washhouse, and let him clean himself with cold water.

Then they took him to the barracks. Fletcher met him at the door, his face creased with concern.

"You look terrible, Roberts."

"Feel worse."

Tom was there too, his expression carefully neutral. Walsh and Miller sat on their bunks across the barracks. They looked at Dick but didn't speak. Their faces said everything. Anger. Resentment. You led us to this.

Dick understood. He deserved that. Deserved worse.

He lay down on his bunk and closed his eyes. His body was broken. His spirit was battered. But underneath it all, underneath the guilt and the exhaustion and the defeat, there was still that small, hard core of determination.

Next time, he'd go alone.

Next time, he'd either make it or die trying.

But there would be a next time. He couldn't stop. Wouldn't stop. Even knowing the cost, even knowing the odds, even knowing what failure meant.

Because some part of him would rather die trying to get home than live accepting that he never would.

Tom sat beside him. Didn't say anything. Just sat there while Dick stared at the ceiling and tried to figure out how to survive one more day in a place that was slowly killing him.

"I'm sorry," Dick said finally.

"I know."

"Walsh and Miller hate me."

"They don't hate you. They're just... they understand now what I already knew. That going with you means suffering. That your determination is contagious and dangerous."

"I won't ask anyone again. Next time I will go alone."

Tom looked at him. "Next time you'll die."

"Maybe. But at least no one else will."

Tom shook his head but didn't argue. What was there to argue? Dick had made his choice. Had made it a long time ago. All Tom could do was watch and wait and hope his friend survived long enough to make it make sense.

Dick pulled out the photograph one more time. Looked

at their faces. His mother. His father. His five sisters. Somewhere in the world, they were alive, living their lives without him. Waiting for him, maybe. Or maybe moving on, accepting that he was gone.

Either way, Dick had to try to get back to them. He had to keep trying. Because the alternative was accepting that he'd never see them again, and Dick couldn't accept that. Wouldn't accept that.

Not yet. Not ever.

He put the photograph away and closed his eyes. Tomorrow he'd start recovering. Building strength. Watching patterns. Planning.

Not today. Today he just needed to survive. To rest. To let his body heal enough to make one more attempt.

Because there would be one more attempt. Dick knew that with absolute certainty.

Next time: alone.

Next time: all the way to freedom or all the way to death.

No middle ground. No half measures. No more guilt over other people's suffering.

Just Dick and the border and whatever lay between.

He'd either make it, or he wouldn't. But at least he'd try. At least he'd know.

And that would have to be enough.

CHAPTER 15 — THE LONG WINTER

Dick returned to the barracks on a grey afternoon in late April. Two weeks in solitary had hollowed him out again, left him weak and dizzy and silent. The other prisoners looked at him briefly, then away. Walsh and Miller didn't look at all.

Tom helped him to his bunk. Brought him soup. Sat with him while Dick stared at the ceiling and said nothing.

"I'm done," Dick said after a long silence. "No more attempts. I'm done."

Tom studied his face. "You mean that?"

"Yes."

Tom wanted to believe him. Dick wanted to believe himself.

Three days later, Dick returned to work detail. The guards assigned him to drainage ditches, the worst work available. Punishment for his second escape attempt, though the commandant hadn't called it that officially. Just permanent assignment to the hardest labour.

The ditch was four feet deep and full of winter's

accumulated mud and debris. Dick climbed down with five other prisoners and began shoveling. The mud was thick and heavy, clinging to the shovel, making each scoop an effort. His wounded shoulder screamed with every lift. His hands, still weak from solitary, barely gripped the handle.

He worked mechanically. Scoop, lift, throw. Scoop, lift, throw. His body moved while his mind went blank. The other prisoners talked quietly among themselves, but Dick didn't join in. Just worked. One shovel at a time. One pile of mud after another.

At midday they stopped for the thin soup and bread ration. Dick sat on the edge of the ditch, his boots caked with mud, and ate without tasting. The bread was sawdust in his mouth. The soup was warm water with bits of turnip. He swallowed it all because his body needed fuel, not because he wanted to eat.

Walsh sat nearby. They didn't speak. Walsh had stopped being angry sometime during Dick's solitary confinement. Now he was just distant, like they'd never been friends at all.

"You look terrible, Roberts," Fletcher said, settling beside him.

Dick shrugged.

"You need to eat more. You're wasting away."

"I'm eating."

"Not enough." Fletcher pushed his own bread ration toward Dick. "Take half. I can spare it."

"No."

"Don't be stupid. Take it."

Dick took the bread and ate it, but tasted nothing. Just chewed and swallowed, chewed and swallowed. Fuel for the body. Nothing more.

After lunch they returned to the ditch. Dick climbed down and resumed shoveling. Scoop, lift, throw. His arms ached. His back ached. His shoulder was on fire. But he kept working because stopping meant drawing attention, and attention meant questions.

The afternoon stretched endlessly. The sun moved across the sky but brought no warmth. Dick's hands developed new blisters on top of old scars. The blisters broke, bled, reformed. He wrapped them with strips torn from his shirt and kept shoveling.

When the guards finally called an end to the work, Dick climbed out of the ditch and stood in line with the others for the march back to camp. His legs trembled with exhaustion. His vision blurred at the edges. But he walked, one foot in front of the other, because that's what bodies did. They kept moving until they couldn't anymore.

Back at the barracks, Tom brought him water. Dick drank without speaking. Lay down on his bunk without removing his muddy boots. Closed his eyes and waited for unconsciousness.

"You're scaring me," Tom said quietly. "This isn't you. This is someone else wearing your face."

Dick didn't answer. He didn't have the energy for words. Just lay there until sleep came.

The days blurred together. Work, eat, sleep. Work, eat, sleep. Each day was identical to the last. Dick stopped marking time in his head, stopped counting days or weeks. What did it matter? One day was the same as another. The number was irrelevant.

His body adapted to the work. His hands developed thick calluses. His back grew stronger. His shoulder never stopped aching but he learned to work through the pain. He became efficient, mechanical, a machine that shoveled mud without complaint.

The other prisoners stopped trying to talk to him. Even Tom gave up after a while, just brought him food and water and sat nearby in silence. Dick appreciated the silence. Words required energy, required caring about communication. Silence required nothing.

June brought heat. The ditches baked in the sun and the mud turned to concrete. Dick swung a pickaxe to break up the ground, swung until his arms were numb, until blisters opened on his palms despite the calluses. Sweat soaked through his thin shirt. The sun burned his neck and arms. But he kept swinging because that's what bodies did.

One evening in July, mail call brought letters for several prisoners. Dick stood in line out of habit, knowing there would be nothing for him. There never was anymore.

"Roberts," the guard called.

Dick looked up, startled. The guard held a single envelope, thin and worn from its long journey.

Dick took it with shaking hands. His mother's handwriting. Posted three months ago. He carried it back to his bunk and sat down, holding it without opening it. Tom watched from across the barracks but said nothing.

Finally, Dick opened the envelope. One page, front and back. His mother's neat script, the words she'd written when he was still someone she recognized.

Dearest Dick,

We received your last letter and are so relieved to know you're alive and as well as can be expected. Your father and I pray for you every night. The girls ask about you constantly. Meryl wants to know when you're coming home. I don't know what to tell her.

Things here are difficult. Rationing is stricter. Your father's hours at the newspaper have increased. Jean is engaged to a boy from her office. Lorna is talking about joining the Women's Auxiliary. The house feels empty without you and your laughter.

I don't know when this war will end or when you'll come home. But we're waiting for you, Dick. However long it takes. Come back to us.

All our love,

Mum

Dick read the letter three times. Then he folded it carefully and put it in his pocket with the photograph. He pulled out the photograph and looked at it. The paper was barely holding together now, the creases worn through in places. The faces were fading. His mother looked younger in the photo than she'd be now. His sisters looked like children, but they were women now. Jean engaged. Lorna wanted to join up but she was working in a factory making uniforms.

He was missing their lives. Missing everything. And there was nothing he could do about it except survive and hope the war ended before too many more years passed.

Dick put the photograph away. Lay down on his bunk. Stared at the ceiling.

Tom climbed up to his own bunk. "Good news from home?"

"My mother's well."

"That's good."

"Jean's engaged."

"That's good too."

Dick said nothing more. Tom fell silent. Eventually they both slept.

The letter should have meant something. Should have reconnected Dick to the world outside this camp, reminded him of what he was surviving for. But it didn't. The words were just marks on paper. The people they described were strangers who happened to share his name.

Dick kept the letter with the photograph and Eileen's old letters. A collection of words from people he used to know. Evidence that he'd once had a life beyond mud and ditches and endless grey days.

By August, the heat was unbearable. Two prisoners collapsed from heat stroke. One died three days later. The guards increased water rations but it wasn't enough. Dick drank until his stomach hurt but was always thirsty. His lips cracked and bled. His skin burned and peeled.

He kept working because stopping meant dying, and

something in him refused to die despite the emptiness. His body insisted on surviving even when his mind had given up.

September brought cooler air and rain. The ditches filled with water that had to be bailed out before work could continue. Dick stood knee-deep in muddy water and scooped it out with a bucket, one load at a time, for hours. His feet went numb. His clothes never dried. He developed a cough that wouldn't leave.

Fletcher died in October.

Dick came back from work detail and found Tom sitting on Fletcher's bunk, his face grey.

"Fletcher?" Dick asked.

"Heart attack. This afternoon. He was gone before anyone could help."

Dick sat down slowly. Fletcher, who'd survived two years in the camp, who'd mentored them all, who'd told Dick to accept his situation and survive. Dead from a heart that just stopped.

"They're burying him tomorrow," Tom said. "Outside the camp. There's a cemetery for prisoners."

Dick nodded. He should feel something. Grief, loss, something. But there was just emptiness where emotions should be.

The next day, a small group attended the burial. Dick stood with Tom and Walsh and Miller and watched them lower Fletcher's body into the ground. The chaplain said words Dick didn't hear. Someone recited a psalm. Then it was over and they returned to camp.

Fletcher was gone. The bunk was empty. Within a week,

a new prisoner would occupy it and Fletcher would be forgotten except by the men who'd known him.

Dick lay in his bunk that night and understood something: this was how it ended for most of them. Not going home. Not escaping. Just dying in camp from disease or exhaustion or hearts that gave up. Fletcher had accepted his situation, had survived day by day, and it hadn't mattered. He'd died anyway, far from home, buried in Polish soil.

Acceptance was just slower dying. Fletcher had been right about survival. But survival wasn't the same as living.

Dick pulled out the photograph one more time. His family. The life he was supposed to return to. But return was becoming more abstract with each passing day. Even if he survived the war, even if he eventually went home, would he still be the person in that photograph? Or would he be someone else, someone shaped by years of ditches and mud and watching friends die?

He didn't know. Didn't know if it mattered.

October turned to November. A year since the first escape attempt with Price. Dick thought about it on the anniversary. The clearing, the snow, the blood. Price falling. That final shot. It felt like something that had happened to someone else, in another life.

November brought the first frost. Then December brought snow and cold that seeped into everything. Dick woke with his blanket frozen to the bunk. His breath steamed in the barracks. The guards brought slightly better rations, not from kindness but because dead prisoners couldn't work.

Christmas came. A few prisoners sang carols quietly. Tom received a parcel from home with tins of food and warm socks. He shared everything with Dick, who accepted without comment and ate without tasting.

"You used to get letters," Tom said on Christmas night. "Parcels. Now nothing."

Dick shrugged. The mail was unreliable. Or his family had stopped writing. Either way, he had no control over it.

"Dick," Tom said carefully. "I need you to hear this. You're disappearing. You're still here, but you're not here. I'm losing you and I don't know how to stop it."

"I'm fine."

"No. You're not fine. You're the opposite of fine." Tom leaned closer. "Fletcher's death should have affected you. The letter from your mother should have affected you. Nothing affects you anymore. That's not survival. That's giving up."

"I'm working. I'm eating. I'm surviving. That's what you wanted."

"I wanted you to live, not just survive. There's a difference."

Dick didn't answer because he didn't know if Tom was right or wrong. He was surviving. His body continued functioning. His heart kept beating. But the part of him that had wanted things, that had fought for things, that had believed in things but that part was gone. Fletcher's death had taken the last of it.

"Promise me something," Tom said. "Promise me you'll hold on until the war ends. Don't give up completely.

Just hold on."

"I promise," Dick said.

It was an easy promise to make because holding on was all he was doing anyway. Just holding on, day after day, with nothing beyond the next meal, the next work shift, the next night's sleep.

The new year came. 1943. Dick had been a prisoner for twenty months. He tried to imagine twenty more months, forty more, sixty more. Tried to imagine being thirty years old and still here, still shoveling ditches, still waiting for a war that never ended.

The numbers made no sense. Time had lost meaning. There was only now: cold, hunger, work, sleep. Repeat until death.

January brought illness. A flu swept through the camp. Dick caught it in the second week, spent five days shivering in his bunk while his body fought fever and chills. Tom brought him soup when there was soup. Sat with him when he was conscious. Covered him with extra blankets when he was delirious.

Dick remembered fragments. Tom's face hovering above him. The taste of broth. Voices speaking German. The sensation of falling through darkness. Then waking weak and soaked with sweat, the fever broken, his body empty.

"You're back," Tom said when Dick opened his eyes.

"How long?"

"Five days. You were out for most of it. I thought..." Tom's voice caught. "I thought you were going to die."

"Should have," Dick said. His throat was raw. "It would

have been easier."

"Don't say that."

But it was true. Dying would have been easier than this. No more ditches. No more cold. No more waiting for a war that had no end. Just sleep and silence and nothing.

Dick recovered slowly. His body took a week to regain enough strength for work. When he returned to the ditches, he was weaker than before, his arms trembling with each shovel load. But he kept working because his body insisted on surviving even when the rest of him didn't care.

February was grey and cold. Dick worked through it in silence. Tom stopped trying to engage him in conversation. Walsh and Miller kept their distance. The other prisoners treated him like a ghost, someone who was there but not really present.

Dick didn't mind. Being a ghost was easier than being human. Ghosts didn't feel. Didn't hope. Didn't wait for things that would never come.

March arrived with the first hints of warmth. Slightly longer days. Ice melting from the barracks roof. Mud instead of frozen ground. The prisoners emerged from winter looking skeletal and grey, but alive.

Dick stood in the parade ground one morning for roll call and realized something: he'd been a prisoner for nearly two years. Twenty-two months. Almost two full years of his life, gone.

He tried to remember what he'd been like before the war. The boy who'd enlisted with his mates, who'd trained in Egypt, who'd thought war would be an adventure. That person was gone completely, replaced by this hollow thing that shoveled mud and ate bread and stared at nothing.

Werner walked past during a work detail. The old guard was moving even slower now, his limp pronounced, his face lined with age and weariness. He stopped and looked at Dick.

"You are not well," Werner said in English.

Dick shrugged.

"You are alive, but you are not living. This is dangerous." Werner paused. "My son, we received word. He is alive. Wounded, but alive in a hospital in Frankfurt. He will come home when he is healed."

"That's good," Dick said. And meant it.

"Yes. It is good. But he is not the same boy who left. His letters, they are different. He writes of things he has seen, things he has done. He has changed." Werner looked at Dick. "You also are changed. War does this. It takes boys and makes them old men."

"Yes."

"But my son still has hope. Still has something to live for. You?" Werner shook his head. "You have given up hope. I see it in your eyes. Empty eyes. Dangerous."

Werner walked away. Dick went back to work.

That night, Dick lay in his bunk and thought about Werner's words. Empty eyes. Is that what he had now? Eyes that saw nothing, felt nothing, wanted nothing?

He pulled out the photograph one more time. His mother. His father. His five sisters. Faces he barely recognized anymore. People who were living their lives while he existed in grey suspension, neither alive nor dead, just waiting.

Somewhere in the world, spring was coming. Trees were budding. Birds were returning. Life was continuing. But here, in this camp, in this barracks, on this bunk, there was just Dick Roberts, ghost prisoner, hollow man, empty eyes.

He put the photograph away and stared at the ceiling. Tomorrow would be the same as today. And the day after that. And the day after that. Until the war ended or he died, whichever came first.

Werner was right. He'd given up hope. Not suddenly, not dramatically. Just slowly eroded it away over months of ditches and cold and watching Fletcher die. Hope had leaked out of him like water from a cracked cup until there was nothing left.

And without hope, what was there? Just existence. Just survival. Just the mechanical repetition of eating and working and sleeping until his body finally gave up too.

Dick closed his eyes and waited for sleep. Tomorrow, another ditch. Another shovel load. Another day of being a ghost.

This was his life now. This was all it would ever be.

And somewhere deep inside, in a place he couldn't quite reach anymore, a small voice whispered that this was wrong. That he wasn't supposed to be a ghost. That giving up hope was the same as dying.

But the voice was distant and fading. Soon it would be gone too. And Dick would be completely empty.

Completely hollow. Completely nothing.

He fell asleep with that thought, and dreamed of nothing at all.

CHAPTER 16 — THE ADVISOR

They came to Dick on a Thursday evening in early April. Three men he'd seen around camp but never spoken to. Younger prisoners, captured more recently, still carrying that restless energy Dick remembered from his own early days.

The tallest one, a Scot named MacLeod, sat on the edge of Dick's bunk. The other two, Anderson and Hughes, stood nearby. Tom was at the other end of the barracks, but Dick saw him notice, saw him go still.

"We need to talk," MacLeod said quietly.

Dick watched them for a moment. The way they stood too close together. The way their eyes kept darting toward the door. The maps they'd been studying when they thought no one was watching.

"Don't," Dick said. "Whatever you're thinking, don't."

"We need your help," MacLeod said.

"I can't help you."

"You've tried to escape. More than once we heard. You made it two weeks once. You know more than anyone here about getting out."

Dick looked at them properly. MacLeod was maybe

twenty-three, tall and lean with red hair and the broad accent of the Highlands. Anderson was shorter, quieter, dark-haired with careful eyes. Hughes was the youngest, probably not even twenty-one, with the soft face of someone who'd barely started shaving before the war came.

"You'll die," Dick said. "Or get caught and wish you'd died. I've been in solitary four times. I've watched a friend die. I've led others into punishment. It's not worth it."

"We're going to try anyway," MacLeod said. "Would you rather we go unprepared?"

It was the same argument Walsh had used. The same logic that had convinced Dick before. And look how that had ended.

"No," Dick said. "Find someone else."

"There is no one else. You're the only one who's survived multiple attempts. You're the only one who knows what works and what doesn't."

"Nothing works. That's what I know."

MacLeod leaned closer. "My brother was killed at Tobruk. My mother's alone in Edinburgh. She's sixty-three years old and working in a munitions factory because her boys are both gone. I need to get home to her." His voice dropped. "I can't sit here for three more years while she ages and dies without me. I can't."

The words hit Dick harder than he expected. Walsh's father dead. MacLeod's mother was alone. All of them were trapped while the people they loved suffered without them.

"If I help you," Dick said slowly, "and you get caught, they'll punish you severely. If you die, that's on my conscience. I already carry enough guilt."

"We're not asking you to come with us," MacLeod said. "We're not asking you to risk anything. Just your knowledge. Just tell us what you learned. Help us have a better chance than you did."

Dick looked at Tom across the barracks. Tom was watching, his face tight with worry. Their eyes met. Tom shook his head once, a small movement. Don't do this.

Dick looked back at the three men. At their desperate faces, their young faces, their faces that still believed escape was possible.

"If I help you," Dick said, "you follow my instructions exactly. No improvising. No taking risks that I don't approve of. Understood?"

"Yes," MacLeod said. Anderson and Hughes nodded.

"And if you get caught, you tell them I knew nothing. You say you overheard me talking about my attempts and planned it yourselves. You never mention my name."

"Agreed."

Dick closed his eyes briefly. He was doing this again. Helping others risk their lives. But what choice did he have? They'd try regardless. At least this way they'd have a chance.

"All right," Dick said. "But we do this carefully. No rushed planning. No mistakes."

Over the next three weeks, Dick became their teacher.

They met in quiet corners, spoke in whispers, made it look like casual conversation while Dick passed on everything he'd learned.

"Eighty miles to the Swiss border," Dick told them, tracing an invisible map in the dirt with his finger. "Southwest. You follow the river valley initially, then cut west through the hill country. Stay off roads. Travel at night when you can. During the day, find cover and rest."

"Food?" Anderson asked. His voice was soft, hesitant.

"You'll have to forage. Farm fields, root vegetables this time of year. Raw if necessary. Steal from barns if you find them isolated. Avoid villages completely. People will turn you in, either from loyalty to Germany or fear of punishment."

"Water?"

"Streams. Rivers. Drink until you're sick of drinking because you don't know when you'll find more."

"Guards?" Hughes asked. The youngest one, the one who looked like he should still be in school.

"Patrols on main roads. Dogs if they track you from camp. German military near the border." Dick paused. "The dogs are the worst. They'll pick up your scent and follow for days. You need to break your trail. Cross streams. Double back. Confuse them."

"How long did you last?" MacLeod asked.

"First time, less than a mile. Second time, fourteen days. Seventy miles. Almost made it to the border."

"What went wrong?"

Dick thought about Miller's ankle, about the fever, about carrying him while the patrols closed in. "One of us got injured. Slowed us down. The dogs caught up."

"So go fast. Don't get injured." MacLeod said it like it was simple.

"Nothing's simple out there. Everything that can go wrong will go wrong. You plan for contingencies, but there are always things you can't predict." Dick met his eyes. "That's why most escape attempts fail. Not lack of planning. Just bad luck."

They absorbed this quietly. Then MacLeod asked, "Why did you try three times? After it failed twice, why try again?"

Dick didn't answer immediately. He thought about his mother's letters, about Eileen's silence, about the photograph disintegrating in his pocket. About the choice between dying slowly in a camp or dying quickly trying to escape.

"Because staying here was worse than dying," Dick said finally. "That's the only reason anyone tries. When you reach the point where you'd rather risk death than spend another day trapped, that's when you go."

The three men looked at each other. Dick saw the recognition in their faces. They'd reached that point.

Over the following days, Dick taught them everything. How to navigate by stars. How to estimate distances. How to identify German patrols by their movement patterns. How to survive on minimal food. How to keep going when your body wants to give up.

He drew maps from memory, marking landmarks

and danger points. He described the terrain in detail, the river crossings, the hill country, the approach to the border. He explained guard routines, work detail schedules, the best time to slip away.

He taught them what had worked and what had failed. The mistakes he'd made. The things Price had done right. The moment Miller's injury had doomed them all.

"You need to be ruthless," Dick said one evening. "If someone gets injured and can't keep pace, you leave them. You don't stop. You don't try to help. You keep moving or you all get caught."

"Leave them?" Hughes looked shocked. "We can't just abandon someone."

"Yes, you can. You have to. That's how you survive." Dick's voice was flat. "I didn't leave my friend. We all got caught because I wouldn't abandon him. Learn from my mistakes."

The three men fell silent, processing this harsh reality.

Tom cornered Dick a few days before their planned attempt. They were alone in the barracks, most prisoners outside for exercise.

"You're really doing this," Tom said.

"Yes."

"After everything. After Price, after Walsh and Miller, after four stints in solitary. You're sending more men out to die."

"I'm giving them information. What they do with it is their choice."

"Don't hide behind that." Tom's voice shook. "Their

blood is on your hands if they fail."

"Yes," Dick said. "I'll take responsibility. I always do."

"Then why do it?"

"Because they're going anyway. With or without my help. At least this way they have a chance."

Tom shook his head. "You're lying to yourself. You can't escape, so you're living through them. Sending them out to do what you won't do."

The words stung because they held the truth. Dick was helping them while staying behind, safe in his acceptance, his surrender. He envied their courage. Hated them for still having hope.

"Maybe," Dick said. "But I'm still helping them."

Tom walked away without another word.

Fletcher appeared in Dick's mind, the conversation they'd had months before. What would Fletcher say about this? Probably that Dick was a fool. Probably that helping others escape while refusing to try himself was backwards thinking. But Fletcher was dead and Dick was still here, still trapped, still hollow.

The next day during work detail, Walsh approached Dick. They'd barely spoken in months, not since Dick had returned from his second solitary confinement.

"I hear you're advising some boys," Walsh said quietly.

"Word travels fast."

"It's a small camp. People notice." Walsh paused. "You helping them the same way you helped me and Miller?"

"Yes."

"That went well for us." Walsh's voice held no anger, just a flat statement of fact.

"I know. I'm sorry."

"Don't be sorry. Be smart. Those boys are going to get caught. You know they will. And if the guards find out you helped, you'll be shot."

"They won't find out. I'm careful."

Walsh studied him for a long moment. "You've changed. Years ago, you would have gone with them. Now you just send others out while you stay safe. Not sure which version of you I respected more."

He walked away before Dick could respond.

The night before the escape, MacLeod, Anderson, and Hughes came to Dick one final time. The barracks was quiet, most men already asleep. The three of them stood by Dick's bunk, their faces pale in the dim light.

"We're going tomorrow," MacLeod said. "During the afternoon work detail. There's a supply delivery scheduled. Guards will be distracted."

Dick nodded. He'd taught them that trick. It's how he'd escaped the second time.

"We wanted to thank you," Anderson said quietly. "For helping us. For giving us a chance."

"Don't thank me yet. Thank me when you're in Switzerland."

"If we make it," MacLeod said, "we'll get word back somehow. Let people know your advice worked."

"And if you don't make it," Dick said, "don't mention my name. Tell them you planned it yourselves."

"We will," Hughes promised. The youngest one, the one who might not survive the first night out there.

They stood awkwardly for a moment. MacLeod extended his hand. Dick shook it. Then Anderson's hand. Then Hughes's. Three men about to attempt something impossible, and the man who'd helped them prepare while refusing to try himself.

"Why don't you come with us?" MacLeod asked suddenly. "You know the route better than we do. You'd increase our odds."

For a moment, Dick considered it. He imagined walking out with them tomorrow, leaving this place behind, trying one more time. The pull was strong, that old desire to be free, to go home, to see his family again.

But then he thought about Price falling in the snow, about Walsh and Miller suffering because of him, about the four times he'd been locked in darkness and cold. About Fletcher dying in his bunk. About becoming a ghost who felt nothing.

"No," Dick said. "I'm done trying. But you three aren't. Go. Make it mean something."

"You could still—" Hughes started.

"No," Dick said again, more firmly. "You go. I stay. That's how it is."

MacLeod looked at him with something like pity. "You're more trapped than any of us, you know that? At least we're still trying. You've already given up."

The words hung in the air. Dick had no answer for them because they were true.

After they left, Dick lay in his bunk and stared at

the ceiling. Tom was silent above him, but Dick knew he was awake, knew he was angry and worried and couldn't understand why Dick kept doing this.

Dick pulled out the photograph, looking at it in the dim light. His mother. His father. His sisters. The life he was supposed to return to. The life that was fading like the photograph itself, becoming less real with each passing day.

Tomorrow, three men would risk everything to get home. And Dick would stay here, safe in his surrender, living through their attempt because he no longer had the courage to make his own.

Maybe MacLeod was right. Maybe Dick was more trapped than any of them. Trapped by fear. Trapped by guilt. Trapped by the weight of failures and deaths and punishments that had broken something inside him.

Fletcher had told him to survive. Dick had survived. But somewhere along the way, surviving had stopped being enough. He'd become a hollow man giving advice to the living, a ghost teaching others how to escape while remaining forever behind.

Dick put the photograph away and closed his eyes. Tomorrow would bring either news of success or the familiar sound of failure. Either way, Dick would still be here, still trapped, still waiting for a war that might never end.

That was his choice. His acceptance. His defeat.

And deep down, in a place he tried not to examine, he hated himself for it.

Tom's voice came from the darkness above. "Dick?"

"Yeah?"

"When they fail tomorrow, and they will fail, what then? Will that finally be enough? Will you finally stop doing this to yourself?"

Dick thought about the question. When MacLeod, Anderson, and Hughes got caught, when they were brought back beaten and broken, would that be the end of it? Would Dick finally accept that escape was impossible, that helping others try was just another form of torture?

"I don't know," Dick said honestly.

"I'm tired," Tom said. "I'm tired of watching you destroy yourself piece by piece. I'm tired of watching you send others out to suffer. I'm tired of being your friend while you refuse to be your own."

"I know."

"But I'm still here. Still sitting in this bunk above you. Still bringing you food when you won't eat. Still trying to keep you alive." Tom's voice broke slightly. "I don't know why anymore. But I am."

"Thank you," Dick whispered.

Tom didn't respond. Eventually, his breathing evened out into sleep.

Dick lay awake for a long time, thinking about tomorrow, about MacLeod and Anderson and Hughes walking into the forest, about the advice he'd given them, about whether it would be enough.

Probably not. Advice was never enough. He'd learned that the hard way.

But at least they'd try. At least they still had that fire, that refusal to accept captivity, that belief that home was possible.

Dick had lost that somewhere along the way. Lost it in ditches and darkness and watching Fletcher die. Lost it in the slow erosion of hope over months of grey days.

The three men would try. Dick would wait. That was the arrangement. That was all he had left.

He fell asleep with that thought, and dreamed of nothing at all.

CHAPTER 17 — THE ATTEMPT

They slipped away on a Thursday afternoon. Dick was working drainage ditches a hundred yards from where MacLeod, Anderson, and Hughes were hauling timber. He didn't watch them go. Didn't want to see the moment they walked into the trees, didn't want that image burned into his memory.

But he heard the absence. That sudden quiet when three sets of footsteps stopped. That pause before the guards noticed. That holding breath of the camp waiting for the alarm.

It came at roll call. In the evening, the sun low and red. The guards counted twice, then a third time. Hartmann's face went rigid. The commandant was summoned.

"Three prisoners are missing," Hartmann announced in German, then English. "MacLeod, Anderson, Hughes. They will be found. They will be punished."

Dick stood at attention, his face neutral, his pulse pounding in his temples. They'd been gone for maybe four hours. With luck, they'd covered ten miles, maybe twelve. The dogs would be sent out soon. The patrols would mobilize. Four hours wasn't enough.

The guards questioned everyone. Has anyone seen anything? Has anyone noticed the missing men acting strangely? Has anyone helped them plan?

Dick was pulled aside with a dozen others, questioned by Hartmann while a guard stood nearby with a rifle.

"You have escaped three times, Roberts. Did you help these men?"

"No."

"You expect me to believe that? You, the camp expert on escape, knew nothing about this?"

"I knew nothing."

Hartmann studied him, suspicious but unable to prove anything. "If I discover you helped them, you'll be shot. Do you understand?"

"Yes."

Dick was released back to the barracks. Tom grabbed his arm, pulled him to their bunks.

"Please tell me you didn't help them," Tom said.

Dick said nothing.

"Christ, Dick. They're going to die out there. And if the guards find out you advised them, you'll die too."

"They won't find out. I was careful."

"Careful?" Tom's voice shook. "You sent three men to their deaths. How is that careful?"

"They chose to go. I just gave them information."

"You're splitting hairs. Their blood is on your hands if they fail."

Dick turned away. Tom was right, of course. Tom was

always right. But it didn't change anything. The men were gone, the attempt was made, and now all Dick could do was wait.

The dogs went out that night. Dick heard them barking in the distance, and heard the shouts of guards organizing search parties. He lay in his bunk and pictured MacLeod, Anderson, and Hughes running through darkness, the sound of dogs getting closer, fear rising in their throats.

Please let them make it, Dick thought. Please let my advice be enough.

But he remembered his own attempts, remembered how nothing had been enough. How bad luck and injury and dogs had always won in the end.

The next day passed in tense silence. The prisoners worked their details without speaking, all of them aware that three men were out there somewhere, running or hiding or already caught. The guards were furious, patrolling constantly, watching everyone with suspicion.

Fletcher's memory came to Dick during the work. What would the old man have said? Probably that Dick was a fool, helping others escape while he stayed behind. Probably that advice without action was cowardice dressed up as wisdom.

Dick dug his shovel into the mud and tried not to think about it.

Tom brought him soup that evening. Sat beside him while Dick ate without tasting.

"What do you think their odds are?" Tom asked.

Dick swallowed the tasteless broth. "Maybe one in ten they make it. Probably less."

"Then why help them?"

"Because one in ten is better than zero."

Tom shook his head. "You're a complicated man, Roberts."

The second day brought no news. The dogs had lost the scent or the men had successfully broken their trail. The guards were baffled, angry. The commandant ordered stricter security, threatened collective punishment if anyone had knowledge they weren't sharing.

Dick kept his face neutral, his mouth shut. But inside he was counting days, counting miles. If MacLeod and the others were still free after two days, they'd have covered maybe thirty miles. Less than halfway to the border, but farther than Dick's first attempt. That was something.

Tom wasn't speaking to him. Walsh and Miller kept their distance, both of them recognizing the pattern: Dick helping others risk what he wouldn't risk himself. The other prisoners watched him with a mixture of respect and wariness. The man who kept trying, even if he only tried through others now.

The third day, Dick let himself believe. Three days meant fifty miles, maybe more. They were past the point where most escape attempts failed. Maybe his advice had been enough. Maybe they'd learned from his mistakes. Maybe they'd actually make it.

He pictured them at the Swiss border, crossing into safety, looking back at Germany one last time before disappearing into freedom. Getting word back somehow that they'd succeeded, that his knowledge had finally helped someone reach home.

It was a good dream. Dick held onto it for three entire days.

On the fourth day, they were brought back.

Dick was in the fields when he heard the commotion. Shouting from the main gate. Guards running. The sound of a wagon arriving. He straightened from his work, his hands tightening on the shovel handle.

They marched all the prisoners to the parade ground. The entire camp, everyone standing in formation while the guards surrounded them with rifles raised. The commandant stood on his platform, his face like stone.

Then the wagon rolled through the gate.

MacLeod was slumped in the back, barely upright. His face was a mass of bruises, purple and black spreading across his cheekbones and jaw. One eye was swollen and completely shut. His lip was split, crusted with dried blood. His clothes were torn and muddy, hanging off him like rags.

Anderson sat beside him, blood still wet from a cut above his eye that ran down his temple and neck, soaking his collar. His hands were bound behind his back, the rope tight enough that his fingers were white. He stared at nothing, his expression blank.

Hughes was lying down in the wagon bed. His leg was twisted at an angle that made Dick's stomach lurch. The bone had broken through the skin just below the knee, white against red. His face was grey, unconscious or dead, Dick couldn't tell from the distance.

"Three prisoners attempted to escape," the commandant announced. "They were caught fifty-seven miles from this camp. They lasted four days. This is what happens to those who try."

The guards dragged the three men from the wagon. MacLeod could barely walk, his legs buckling. Anderson stumbled, nearly fell. Hughes had to be carried, two guards holding him under the arms while his broken leg dragged behind, leaving a trail in the dirt.

They were lined up before the assembled prisoners. MacLeod lifted his head, his one good eye searching the crowd. He found Dick in the rows of men. Their eyes met for just a moment. MacLeod's expression held no blame, just exhaustion and defeat and something else, a question, maybe. Was it worth it?

Dick had no answer.

"These men will spend three weeks in solitary confinement," the commandant continued. "After that, they will be assigned to the hardest labour available. Let this be a lesson to all of you. Escape is impossible. Attempting it brings only suffering."

The guards dragged the three men toward the punishment block. Hughes left a streak of blood across the parade ground. Dick watched them disappear through the doorway, then the door closed and they were gone.

After the prisoners were dismissed, Dick returned to work. His hands moved the shovel mechanically while his mind replayed the scene. MacLeod's bruised face, that split lip, that swollen eye. Anderson was bleeding down his neck. Hughes's bone breaking through skin.

Three men broken because Dick had given them hope.

Tom found him that evening.

"You saw?" Tom asked.

"Yes."

"Fifty-seven miles. That's actually impressive. Your advice worked, at least partially."

"Not well enough. They still got caught."

"They were always going to get caught. You knew that." Tom paused. "Does it change anything for you? Seeing them brought back like that?"

Dick looked down at his scarred hands, at the mud under his nails, at the blisters that never quite healed. He pictured MacLeod's face, that one eye finding him in the crowd. He pictured the question in that look.

"Yes," Dick said slowly. "It changes things."

"How?"

Dick looked at Tom, his friend who'd been with him through everything, who'd tried so hard to keep him alive, who'd watched him fail again and again.

"I think I need to try again," Dick said.

Tom went very still. "What?"

"I need to try one more time. Myself. Not through others. Myself."

"Are you insane? You just watched three men get caught. You've been caught three times yourself. What makes you think a fourth attempt would be any different?"

Dick pulled out the photograph, the paper barely

holding together now. "Because fifty-seven miles proves it's possible to get that far. Because my second attempt made it seventy miles. Because I know things now that I didn't know before. Because I've learned from every failure."

"You've also been punished for every failure. One more attempt and they'll shoot you, Dick. The commandant made that clear."

"Maybe. Or maybe the Germans have stopped caring. It's been a year since my last attempt. They might not even be watching me anymore. They might think I've given up."

Tom grabbed Dick's shoulders. "Listen to me. You have given up. For a year, you've been quiet and obedient and accepted your situation. That's good. That's healthy. Don't throw that away now."

"I wasn't giving up. I was waiting." Dick met Tom's eyes. "Waiting for the guards to forget about me. Waiting for their vigilance to drop. Waiting for the right moment."

"You're rationalizing. You're watching those three men fail and somehow turning it into motivation to try again. That's not logical. That's an obsession."

Dick shook his head. "It's not an obsession. It's understanding. They made it fifty-seven miles with only my advice. I've made it seventy miles with experience. I speak some German now. I know the terrain. I know what works and what doesn't. I can do this."

"You can't. No one can. The border is eighty miles away through enemy territory. The odds are impossible."

"One in ten," Dick said. "That's still a chance. One in ten is better than zero."

"And nine in ten means death or capture. You're betting your life on a ten percent chance."

"I've been doing that since Crete. We all have." Dick put the photograph away. "I'm going to try again, Tom. One final time. I'm going alone. I'm telling no one when or how. I'm not asking for help. I'm just doing it."

Tom stared at him for a long moment. Then he let go and stepped back.

"I can't watch this again. I can't watch you throw yourself against that wall one more time knowing you'll fail."

"Then don't watch. But I'm doing it anyway."

Tom walked away. Dick stood alone in the barracks, his decision made.

He'd been broken for a year. Hollowed out. Accepting. Surrendered. But watching MacLeod, Anderson, and Hughes being dragged back to punishment had reminded him of something he'd forgotten: trying and failing was still better than not trying at all.

Those three men had made it fifty-seven miles. They'd lasted four days. They'd tried. And yes, they'd failed. Yes, they'd be punished. But they'd tried.

Dick had stopped trying. Had spent a year playing dead, convincing the guards he'd given up, convincing his fellow prisoners he'd accepted his fate. Maybe even convincing himself.

But he hadn't. Not really. He'd just been waiting. Letting time pass. Letting the guards forget about him. Letting their vigilance fade.

And now, watching those three men suffer for their attempt, Dick understood: his advice had gotten them fifty-seven miles. His experience had taught them well enough to beat his first attempt. They'd failed, but they'd done better than he'd expected.

Which meant if Dick tried again, with all his knowledge, all his experience, all the lessons learned from four years as a prisoner, he might just succeed.

The odds were still terrible. Nine in ten chance of failure. But that tenth chance existed. That tiny possibility of freedom. Of home. Of seeing his family again.

Dick had spent a year believing that possibility was gone. Now, watching three men be dragged to punishment, he believed again.

Maybe Tom was right. Maybe this was obsession, not logic. Maybe Dick was turning failure into motivation because he couldn't accept the alternative.

But Dick didn't care. He was done accepting. Done waiting. Done being hollow.

He was going to try one more time. And this time, he'd either make it all the way to Switzerland, or he'd die trying.

Those were the only two options left. Freedom or death. Nothing in between.

Dick pulled out the photograph one final time. He looked at the faces. Made his silent promise: I'm coming home. One way or another, I'm coming home.

Then he put it away and began to plan.

The guards thought he'd given up. His fellow prisoners thought he'd accepted his fate. Even Tom thought he'd surrendered.

Let them think it. Let them believe the lie. Because when Dick finally made his move, when he finally walked away from this camp one last time, no one would see it coming.

And that, Dick realized, was his advantage. After a year of playing dead, he'd become invisible. The guards didn't watch him anymore. The commandant had stopped caring. Werner barely noticed him.

Dick had disappeared into the background of camp life. Become just another prisoner, serving his time, waiting for the war to end.

Perfect.

Because invisible men could escape. Invisible men could slip away unnoticed. Invisible men had a chance.

Dick didn't know when the opportunity would come. Didn't know how long he'd have to wait. But he'd wait as long as necessary. He'd been waiting for a year already. He could wait a few more weeks, a few more months if needed.

And when the moment came, when everything aligned, Dick would go. Alone. Silent. Final.

One last attempt. Freedom or death.

Dick was ready.

Walsh appeared beside him as the prisoners settled in for the evening. Sat on the edge of Dick's bunk without invitation.

"Heard you're planning something," Walsh said quietly.

"Maybe."

"After watching those three get dragged back? You're crazier than I thought."

"Maybe."

Walsh studied him for a long moment. "You're different than you were. A year ago, you were desperate, reckless. Now you're cold. Calculating. Not sure which version scares me more."

"I'm not asking you to come," Dick said.

"I know. You won't ask anyone this time. You'll go alone." Walsh paused. "Good. That's smart. We all failed because we went in groups. Maybe solo is the answer."

"Maybe."

"When?" Walsh asked.

"When the time's right. Could be weeks. Could be months. I'll know when I see it."

Walsh nodded. Stood to leave. Then stopped. "Dick? If you make it, tell my mother I tried. Tell her I didn't just sit here waiting to die. Tell her I fought."

"I will," Dick said.

Walsh walked away. Dick lay back on his bunk and stared at the ceiling.

The decision was made. The plan was beginning. Not tonight. Not tomorrow. But soon.

Dick closed his eyes and, for the first time in a year, allowed himself to imagine freedom. The Swiss border. Safety. Home.

It might be impossible. Probably was impossible. But the impossible was better than nothing.

And nothing was all Dick had been for the past year.

Time to be something again. Time to try.

One last time.

CHAPTER 18 — THE FINAL ESCAPE

Dick came out of solitary on a Thursday in early June. The guards unlocked the door and he walked into sunlight that stabbed his eyes, into air that felt too vast after three weeks in a cell barely bigger than a coffin. His legs trembled. His shoulder throbbed with that deep ache that had become permanent.

Tom met him at the barracks door. Neither spoke. Tom just gripped his good shoulder briefly, his face taut.

"I'm done," Dick said. "I'm not trying again."

Tom searched his face. "You mean that?"

"Yes."

Tom nodded slowly, wanting to believe it. Dick wanted to believe it too.

But even as he said the words, he was already planning.

The days stretched into weeks. Dick was assigned to the hardest labour details as punishment. Digging latrines. Hauling coal. Breaking rocks. Work designed to break the body and the spirit. His wounded shoulder screamed with every movement, but he gritted his teeth and kept working because stopping meant more punishment.

Fletcher's memory haunted him during the work. The old man had survived by accepting his cage, and had died anyway, buried in Polish soil far from home. Dick shoveled coal and understood: acceptance was just slower dying.

But Dick went through the motions anyway. Roll call. Work. Meals. Sleep. Each day was identical to the last. The summer heat built, making the work brutal. Men collapsed from exhaustion. Dick kept going, his body moving on autopilot while his mind worked elsewhere.

He watched the guards. Not obviously, not like before when his intentions had been written on his face. This time he was careful. He memorized Werner's patterns, noted which guards were alert and which were lazy, observed when supplies arrived and when security was thinnest.

At night, lying on his bunk with Tom snoring above him, Dick would pull out the photograph. The blood stains from his wound had faded to brown. The paper was like tissue now, threatening to tear with each unfolding. His family looked back at him from another life.

July came. The war news filtering into camp suggested the Allies had invaded Italy, that the Germans were being pushed back in Russia. But turning tides didn't help prisoners. They were still here, still trapped, still waiting.

Dick couldn't wait. Something in him had crystallized during those weeks watching MacLeod, Anderson, and Hughes being dragged back bleeding. He was going to try again. One more time. And this time he'd either make it or die trying.

Tom noticed, of course. Tom always noticed.

"You promised," Tom said one evening. They were alone in the barracks, most of the others at an impromptu concert.

"I did."

"You're planning it anyway."

Dick didn't deny it.

"They'll kill you this time. The commandant made it clear. Another attempt means execution."

"I know."

Tom's jaw tightened. "Why? Why can't you just accept this and wait?"

Dick looked at his friend, at the man who'd been with him since Egypt, who'd tried so hard to keep him alive.

"Because waiting is its own kind of death," Dick said. "Because every day here is a day I'm not with them." He pulled out the photograph. "Because if I wait until the war ends, I'll be someone who learned to accept cages. And I can't be that person."

Tom looked away. "When?"

"I don't know yet. When the opportunity comes."

Tom was quiet for a long moment. Then he pulled something from his pocket and pressed it into Dick's hand. A map, hand drawn, showing the route from the camp to the Swiss border.

"Miller drew it before the second escape. He never gave

it to you then." Tom's voice was rough. "I kept it. Don't know why."

Dick stared at the map, the careful lines, the landmarks marked.

"If you're going to do this," Tom said, "you have to make it. You understand? Price died. You've been shot. You've spent weeks in solitary. All of that has to mean something. So don't just try. Do it. Make it all the way. Or don't go at all."

Dick folded the map carefully and tucked it into his shirt. "I'll make it."

"Swear it."

"I swear."

The opportunity came on a grey afternoon in late August. A supply train had derailed five miles north. All the younger guards were sent to help with salvage. Only Werner remained, supervising a small work detail near the perimeter.

Dick was moving timber from the saw pit to the storage shed, repetitive work requiring minimal supervision. Werner stood in the shade of the guardhouse, pipe in hand, attention elsewhere. The other prisoners were scattered across the work area.

Dick set down his load of timber. He looked around. No one was watching. Werner turned away. The gate to the service road was visible maybe fifty yards away. Beyond that, forest.

Now.

Dick walked away from the timber pile, moving with purpose but not urgency. He reached the gate, opened it as if he had every right to be there, and walked through. No shouts. No alarms. He kept walking, picking up speed gradually, heading for the tree line.

He reached the forest and looked back once. The camp sat there, grey and quiet. Werner was still by the guardhouse, unaware.

Dick turned and ran.

He ran until his lungs burned and his legs threatened to give out, until the camp was far behind and the forest thick around him. Then he slowed to a fast walk, conserving energy, putting distance between himself and pursuit.

They'd discover him missing at evening roll call. Maybe three hours. Three hours to get as far as possible before the dogs came.

Dick pulled out Miller's map. The Swiss border was approximately eighty miles southwest. Between here and there was enemy territory, German patrols, civilians, rivers and mountains.

He followed deer trails through the forest, staying off roads, avoiding open ground. His shoulder ached with every step. His body was weak from months of inadequate food and hard labour. But he kept moving, one foot in front of the other.

Night fell. Dick kept walking, using the stars to navigate. The temperature dropped and he had no coat, only his thin prison uniform. He shivered but kept moving.

Around midnight, dogs barked in the distance. They'd picked up his trail. Dick splashed through a stream to break his scent, then changed direction, angling west before turning south again.

The barking faded. He'd bought himself time.

By dawn, Dick was exhausted and starving. He found a dense thicket and crawled into it, pulling branches over himself for concealment. Then his body gave out despite his fear, despite the dogs.

He woke to afternoon light and voices. German voices, close. Dick lay perfectly still, barely breathing. Through the branches he saw them, two soldiers walking along a trail, rifles slung casually, talking about leave rotations. They passed within ten feet without seeing him.

Dick waited until their voices faded before moving. His body was stiff. His stomach was cramped with hunger. He'd had nothing to eat since the morning bread ration at camp, almost thirty hours ago.

He continued south, following the river shown on Miller's map. Late afternoon, he found a small field with potato plants. He dug with his hands, pulling up four potatoes, brushing off dirt, and eating them raw. They were hard and earthy and tasted better than any meal he remembered.

Water he got from a stream, drinking until his stomach hurt.

The second night was colder. Dick's teeth chattered violently. He found a hollow tree and curled up inside it, hugging himself. Sleep came in brief snatches between shivers.

By the third day, Dick's mind was starting to slip. The lack of food, the cold, the exhaustion, all of it breaking down his defenses. He saw Price walking ahead through the trees. He saw his mother standing by the stream. He saw Eileen in a clearing, her arms outstretched, but when he reached for her she vanished like smoke.

He came to a road and had to cross. He waited until dark, watching for patrols, then ran across and dove into the trees on the other side. No shouts. No shots.

On the fourth day, weak and dizzy, Dick reached a farm. It sat in a small valley, isolated from other buildings. Smoke rose from the chimney. A man and woman were working outside a barn, the man loading hay onto a cart, the woman hanging washing on a line.

Dick watched from the trees. Approaching them was dangerous. They could turn him in. He could be shot.

But he was dying out here. Without food and warmer clothes, he wouldn't make it another day.

Dick stood and walked slowly into the open, his hands raised, showing he was unarmed. The man saw him first and stopped working, his body going tense. The woman turned and her hand flew to her mouth.

Dick stopped twenty feet away. "Australian," he said in German. It was the only word he could think of that might help. "Prisoner. Escaped."

The man and woman looked at each other. Silent conversation passed between them. Then the woman motioned for him to come closer.

Dick approached slowly. The man was perhaps fifty, weathered from outdoor work. The woman was younger, maybe forty, with kind eyes and work-worn hands.

"You are escaped?" the man asked in German.

"Yes."

"They will shoot you if they catch you."

"I know."

"They will shoot us too, for helping you."

"I know. I'm sorry. I need food. And clothes. Then I'll go."

The woman spoke rapidly to her husband in German too quickly for Dick to follow. The man shook his head. The woman persisted. Finally, he sighed and nodded.

"Come," he said to Dick. "Quickly. Into the barn."

They led him inside. The barn was warm and smelled of hay and animals. The woman disappeared and returned with food. Bread and butter. Cheese. Cold meat. Milk. Dick ate slowly, forcing himself not to gulp it down.

"What is your name?" the woman asked in careful English.

"James. But everyone calls me Dick."

"I am Anna. This is my husband, Karl." She paused. "We have a son. He is a soldier in Russia. We do not know if he is alive."

Dick understood then. They were helping him because of their son. Because somewhere, maybe, someone was helping their son. Or maybe their son was dead, and helping Dick was the only way to honor his memory.

"Thank you," Dick said. "I won't forget this."

Karl left and returned with clothes. Old farm trousers, a wool shirt, a heavy coat, and a cap. "Your uniform, we must hide it. Bury it. No one can know you were here."

Dick changed in the corner of the barn, his prison uniform discarded like a shed skin. The farm clothes were warm and clean. Karl buried the uniform behind the barn while Anna brought more food.

"You rest today," Anna said. "Tonight, too dangerous to move. Tomorrow you go."

Dick wanted to protest, wanted to keep moving. But his body was collapsing. One day. He could risk one day.

He slept in the hay loft, real sleep for the first time since escaping. When he woke, it was dark. Karl brought him dinner, hot soup with vegetables and pieces of meat. Dick ate slowly, his stomach unused to richness.

"The border," Dick said. "How far?"

Karl pulled out a rough map. "Maybe fifty kilometers. Three days walking if you are strong. You are not strong. Maybe five days. You follow the river south until it turns east. Then you go west over the hills. The border is in the valley beyond. There is a fence, but it is old. You can cross."

"Are there patrols?"

"Yes. German soldiers near the border. You must be very careful. They shoot on sight." Karl paused. "Many try to cross. Not many make it."

Dick absorbed this. The odds were terrible. But he'd come this far.

He stayed at the farm for two weeks, not one day. He was simply too weak to leave. Karl and Anna fed him, let him rest, helped his body recover from months of starvation and abuse. Dick helped with chores when he could: mending fences, stacking hay, milking the cow.

Anything to repay their kindness.

His body slowly came back to life. The shaking stopped. His strength returned. The constant gnawing hunger finally eased.

At night, they listened to the wireless. Reports of Allied bombing raids. News that the eastern front was collapsing. Karl would translate the important parts, his weathered face growing more hopeful with each broadcast.

"The war is ending," Karl said one night. "Not this year. But maybe next year. Maybe the year after. Germany is losing."

"Your son?" Dick asked.

Karl shook his head. "No word. Six months now."

Anna touched her husband's hand. They sat in silence, three people bound by war's weight, by loss, by hoping for things that might never come.

Dick worked in the fields during those two weeks, helping Karl harvest late potatoes. The physical labour was different now. Not punishment, but purpose. His hands moved through soil, pulling up vegetables that would feed this family through winter. The sun warmed his back. The air smelled of earth and growing things.

One afternoon, Anna brought water to the field. She sat on the stone wall and watched Dick work.

"You have family?" she asked.

"Yes. My mother and father. Five sisters. A girl I was going to marry."

"She waits for you?"

"I don't know. Haven't had a letter in months. She might think I'm dead. Might have moved on."

Anna nodded. "War makes everything uncertain. My Karl, when he was in the first war, I waited four years. Four years not knowing. But I waited." She smiled. "He came home. We got married. We had our son. The waiting was worth it."

"And now your son is in Russia and you're waiting again."

"Yes. We will wait again." Her smile faded. "But this time, I do not know if God will be so kind."

They sat in silence, watching clouds move across the sky.

"You must go home to your girl," Anna said finally. "You must make the waiting worth something."

On the fourteenth day, Karl came to Dick in the barn. His face was tight.

"Tomorrow you must go. Frau Schneider saw you yesterday, working in the field. She asks questions. It is not safe for you here anymore."

That evening, Anna pressed a rucksack into his hands. Food for several days. A water bottle. A blanket. A small knife. "For your journey."

"I can't repay you," Dick said.

"You repay us by surviving," Anna replied. "You repay us by going home to your family. You tell them about us, yes? You tell them not all Germans are monsters."

"I will. I promise."

Karl shook his hand, his grip firm and calloused. "Follow the river. Be careful at the border. The guards there, they shoot first. You understand?"

"I understand."

Anna hugged him briefly. When she pulled back, her eyes were wet. "Go with God."

Dick left at first light, the rucksack on his back, his body stronger than it had been in months, his heart heavy with gratitude and guilt. Karl and Anna had risked everything. If he was caught, if authorities traced him back to their farm, they'd be executed. And they'd done it anyway, because their son was somewhere in Russia and maybe someone was helping him too.

Dick followed the river south, moving carefully, avoiding roads and villages. The map Karl had given him was better than Miller's, showing landmarks and safe routes. Dick studied it constantly, burning every detail into memory.

The days passed by walking. His feet blistered inside the old boots Karl had given him. His shoulder ached. But the food Anna had packed kept him going. He rationed it carefully, eating just enough to maintain strength.

On the third day from the farm, he had to hide from a German patrol. Four soldiers walking along the road, their rifles ready, their eyes scanning the forest edge. Dick pressed himself into a ditch, face down in dead leaves, and didn't move until their boots stopped echoing on the packed dirt.

On the fifth day, he saw the fence. It ran along a ridge in the distance, a dark line against grey sky. The border. Switzerland. Freedom.

Dick stopped and stared. He was close. So close. But Karl had warned him: the last stretch was the most dangerous. Guards patrolled regularly. Orders to shoot anyone trying to cross.

Dick waited until dark. Then he approached the fence, moving slowly, using every bit of cover. The fence was old barbed wire, sagging in places, rusted. Beyond it, neutral territory.

Dick found a spot where the wire was lowest. He lay flat and crawled under, the barbs catching his coat but not stopping him. His hands touched soil on the other side. Swiss soil. Free soil.

He stood on shaking legs. Behind him, Germany. Ahead, Switzerland.

He'd made it.

But he couldn't celebrate yet. He needed to find authorities, needed to turn himself in to Swiss officials before German patrols saw him and didn't care which side of the border he was on.

Dick walked through the night, following a road now, not hiding anymore. Near dawn, he saw a farmhouse with a red tile roof. A Swiss flag hung from a pole.

Dick approached the door and knocked. An older man opened it, looked at Dick's disheveled appearance, and said something in French.

"English?" Dick asked. His voice was hoarse from disuse.

The man switched languages. "You are English?"

"Australian. I escaped from a German prison camp. I need help."

The man's eyes widened. "Come inside. Quickly."

Dick stumbled through the door. His legs gave out and he sat heavily on a chair. The man's wife appeared, took one look at Dick, and started gathering food and blankets.

"You are safe now," the man said. "You are in Switzerland. The Germans cannot touch you here."

Dick nodded. He tried to speak but his voice wouldn't work. Tears ran down his face, hot and unexpected. He'd made it. After three escape attempts, after being shot, after weeks in solitary, after Price's death and Tom's fears and everyone telling him it was impossible.

He pulled out the photograph with shaking hands. It was barely holding together, the creases worn through in places, the blood stains faded but visible. His family looked back at him. His mother, his father, his five sisters.

"I'm coming home," Dick whispered to their faces. "I'm actually coming home."

The Swiss farmer and his wife looked at each other. They'd seen escapees before, had seen men broken by war stumble across the border desperate for safety.

"You rest now," the wife said in accented English. "You are safe. You are free."

Free.

The word settled over Dick like absolution. He was free. The war wasn't over, the camp still existed, Tom and Walsh and Miller and all the others were still there. But Dick was out.

He pictured Tom giving him the map. Werner looking

the other way. Karl and Anna risking everything. Price falling in the snow, dying for the same dream of home that had kept Dick going.

All the cost and sacrifice and luck and timing that had brought him to this moment.

The Swiss farmer's wife brought soup and bread. Dick ate slowly, his stomach no longer accustomed to richness. The warmth spread through him, chasing away the cold that had lived in his bones for so long.

"What happens now?" Dick asked.

"We take you to authorities," the farmer said. "They will process you. Then, eventually, you go home. To Australia."

Home. Australia. The words seemed impossible. Dick had been away for three years. Three years of war and suffering and survival. Three years of becoming someone different.

Would Eileen recognize him? Would his family? Would he recognize himself?

Those were questions for later. For now, Dick was simply alive. Free. With a chance at home however distant it still seemed.

The farmer showed him to a small room with a real bed. Dick lay down and closed his eyes. For the first time in three years, he slept without fear. Without the cold. Without certainty that tomorrow would bring only more suffering.

He slept and dreamed of Newcastle. Of the beach and the house in Alfred Street and his mother's kitchen. He dreamed of Eileen's face and Tom's voice and all the things he'd thought he'd lost forever.

And when he woke, he was still free. Still safe. Still alive.

The war would continue. The camp would continue. Tom and the others would continue to wait and hope and survive. But Dick was out. Dick had made it.

And someday, however long it took, he'd make it all the way home.

CHAPTER 19 — THE LONG ROAD HOME

The Swiss officials processed Dick at a facility in Geneva, a converted hotel with clean white walls and windows that looked out onto Lake Geneva. After three years of dirt and cold and hunger, the cleanliness felt almost aggressive. Everything was too bright, too orderly, too much.

A Red Cross nurse checked his vitals and made notes on a clipboard. She spoke English with a French accent, her voice gentle as she asked questions about his health. Dick answered mechanically. Yes, he'd been shot. Yes, he'd had dysentery. Yes, he'd lost weight. The questions felt distant, as if they were about someone else's body.

"You are severely malnourished," the nurse said. "We will start you on supplemental meals. Small portions at first. Your stomach needs time to adjust."

Dick nodded. He'd gained some weight during his two weeks at Karl and Anna's farm, but he was still gaunt, his clothes hanging loose on his frame. The nurse weighed him and wrote down the number. Dick didn't ask what it was. He didn't want to know.

They gave him a room with two other escaped prisoners. A South African named De Wet who'd gotten out of a camp near Munich, and a British pilot named Patterson who'd been shot down over France and spent two years on the run. The three of them barely spoke. They were all still adjusting to the simple fact of freedom, of being able to walk through a door without permission, of sleeping without guards.

At night, Dick would lie awake listening to the others. De Wet had nightmares, shouting in his sleep in Afrikaans. Patterson paced the room, unable to settle. Dick woke up gasping most nights, convinced he was back in the punishment cell, or running through the snow while shots cracked around him, or watching Price fall.

During the day, they attended briefings. Red Cross officials explained the process. They'd remain in Switzerland until transport could be arranged to England. From there, they'd be processed again, given medical attention, then sent home to their respective countries. It could take weeks or months. No one could say for certain.

"Just more waiting," De Wet said one evening. They were sitting in the common room, drinking real coffee, something Dick hadn't tasted in three years. "We escaped the camps only to wait in Switzerland."

"At least we're not starving," Patterson said.

"True enough."

Dick looked out the window at the lake. Boats moved across the water, their white sails bright against the blue. Normal life, continuing as if there were no war. Switzerland had maintained its neutrality, and had stayed out of the madness. These people hadn't experienced what Dick had experienced. They didn't know what it was like to be hunted, to be shot, to watch friends die.

Resentment flashed through him, sharp and unexpected. Then he pushed it away. It wasn't their fault. They were lucky, that's all. Lucky to be born in a country that wasn't being torn apart.

Dick wrote a letter to his mother, his hand shaking as he formed the words.

Dear Mum,

I'm in Switzerland. I'm safe. I escaped from the camp and made it across the border. I'm being processed by the Red Cross. They say I'll be sent to England soon, then home to Australia.

I'm all right. Thinner than I should be, but they're feeding me well. The doctors say I'll recover.

I don't know how long it will take to get home. Weeks or months. But I'm coming. Tell everyone I'm coming.

All my love,

Dick

He didn't mention the three escape attempts. Didn't mention being shot. Didn't mention Price's death or the weeks in solitary. Those were things he'd tell her in person, maybe. Or maybe never. Some things were too heavy to put in a letter.

He gave the letter to the Red Cross to mail, then went back to his room and pulled out the photograph. It was barely holding together now. The blood stains had faded to brown. The creases were worn through in places. But the faces were still visible. His mother, his father, his five sisters. All of them were waiting for him.

Dick had carried this photograph through everything. Maleme. The withdrawal. The march to the transit camp. The journey to Poland. The farm. The escape attempts. The punishment cells. Every time he'd thought about giving up, about letting go, he'd pulled out this photograph and remembered why he had to survive.

Now he was free. Now he was going home. The photograph had done its job.

Three weeks passed. Dick gained weight slowly, his body adjusting to real food. The nurses monitored him carefully, checking for signs of refeeding syndrome, a dangerous condition that could kill starving men who ate too much too quickly. Dick forced himself to eat slowly, to be patient, even though his body screamed for more.

He walked by the lake every day, just to prove he could. Just to feel the sun on his face and know that no one would stop him, no one would shoot him, no one would drag him back to a cell. Freedom was still strange, still unreal. Part of him expected to wake and find this was all a dream, that he was still in the punishment block at Stalag VIII-A.

De Wet was transferred first, sent to England with a group of South Africans. Patterson left a week later. Dick remained, waiting for his turn. More escaped prisoners arrived. Americans who'd gotten out of camps in Germany. French resistance fighters who'd been imprisoned and escaped. A Norwegian who'd walked across two countries to reach Switzerland.

They all had stories. They all had scars, visible and invisible. They all had the same haunted look, the same difficulty believing they were truly free.

Finally, in late November, Dick's name was called. He was being transferred to England with a group of British and Commonwealth prisoners. They'd leave the next morning.

That night, Dick packed his small bag. The farm clothes Karl had given him. The knife Anna had pressed into his hands. The photograph. The letters from home, worn soft from reading. That was everything he owned in the world.

The journey to England took three days. First by train through Switzerland and France, through countryside that still bore the scars of war. Bombed buildings. Destroyed bridges. Fields turned to mud. Dick stared out the window, watching Europe's wounds pass by.

Then by boat across the Channel to Southampton. The grey water, the grey sky, the grey cliffs rising as they approached England. Dick stood on deck and remembered the last time he'd seen these waters, sailing away to war in 1940. He'd been young then. Naive. He'd thought war would be an adventure.

Now he was coming back broken and scarred, having learned that war was just suffering and survival and watching good men die for nothing.

They docked at Southampton and were taken to a reception camp outside the city. Rows of barracks, much like the camps Dick had lived in as a prisoner, but cleaner and with British flags flying. The irony wasn't lost on him.

Medical examinations came first. A doctor checked Dick's weight, his blood pressure, and his shoulder. The doctor prodded the scar tissue where the bullet had torn through and made notes.

"Permanent damage to the rotator cuff," the doctor said. "Limited range of motion. You'll never have full use of that shoulder again."

"I know."

"Does it cause pain?"

"Sometimes."

"I'll note it in your file. You may be eligible for a disability pension."

Dick nodded. He didn't care about pensions. He just wanted to go home.

After the medical exam came debriefing. Military intelligence wanted to know everything about the camps. Where he'd been held. How many prisoners. Guard routines. Treatment of prisoners. Escape attempts.

Dick told them about Stalag VIII-A, about the work brigade, about the three escape attempts. He told them about Price dying in the snow. About the punishment cells. About being shot. The intelligence officer wrote it all down without expression.

"You mentioned other prisoners," the officer said. "Names?"

Dick gave them what he remembered. Tom Wilson. Walsh. Miller. Fletcher. Cooper. All of them were still in the camp when Dick escaped. All of them are still prisoners.

"We'll add them to our records," the officer said. "As camps are liberated, we're tracking survivors. If your friends are alive, we'll let you know."

After the debriefing, Dick was given a bunk in one of the barracks and told to wait. A ship to Australia would be arranged, but it could take weeks. Everything took time. Everything required patience. More waiting, just like De Wet had said.

Dick settled into the routine of the reception camp. Meals three times a day, medical checks, briefings about what to expect when they returned home. The army was trying to prepare them for civilian life, but nothing could truly prepare a man for going home after years of war.

On his third day in England, mail call was announced. Dick hadn't expected anything, he'd only just arrived, but when his name was called, his pulse quickened.

The corporal handed him an envelope. His mother's handwriting. The letter had been chasing him across Europe, forwarded through Red Cross channels from Switzerland to France to England. Dick took it to a quiet corner of the barracks and opened it with shaking hands.

> *Dear Dick,*
>
> *We received word from the Red Cross that you escaped and reached Switzerland. The relief we felt was beyond words. We had feared you dead after hearing nothing for so long. To know you are alive, that you are safe, that you are coming home is more than we dared hope.*
>
> *The house at Alfred Street was sold last year.*

We moved to a larger place in Broadmeadow, 132 Brunker Road. You'll like it, I think. More room for everyone. A big backyard. Your father has a shorter walk to the paper.

Jean married last month. A man named Alistair who works at the steelworks. They're very happy. Lorna is still arguing with everyone, but she's mellowed somewhat with age and she has met an airman named Keith. Dorrie is training to be a teacher. Shirley and Meryl are growing so fast. Meryl is almost as tall as I am now.

Dick smiled at the image. His baby sister, almost grown. Jean married. All of them aging, changing, living their lives while he'd been frozen in captivity.

Then he reached the next paragraph and the smile faded.

Darling, there is something I must tell you before you come home. Eileen Browning married last year. He is a doctor at the hospital, a good man. They have a baby son now, born in August. I wanted you to hear this from me rather than as a surprise when you arrive.

She waited as long as she could. She asked after you every time I saw her at the shops. But three years is a long time, and none of us knew when the war would end or if you would make it home. When Dr Harrison proposed, she said yes. I hope you can understand. We are all just doing our best to survive this war.

Your sisters send their love. Your father is eager to see you. We are all counting the days until you walk through our door.

Come home safe, my darling boy.

Mum

Dick read the paragraph about Eileen three times. Then he folded the letter carefully and put it in his pocket beside the photograph. He stood and walked out of the barracks, past the other men, past the guards at the gate who nodded him through, out into the grey English afternoon.

It was raining, a fine mist that soaked through his clothes within minutes. Dick walked until he found a bench overlooking a field, sat down, and let the rain wash over him.

Eileen was married. Had a baby. Had chosen life over waiting for a ghost.

Dick sat with it. The loss. The finality. He couldn't summon anger or betrayal. Three years was a long time. Three years of not knowing if he was alive or dead, if he'd ever come home, if waiting was worth it. She'd made the choice that made sense. She'd chosen certainty over hope.

He pulled out the photograph, protecting it from the rain with his hand. His family looked back at him. Eileen had never been in this picture. She'd existed only in his letters, in his hopes, in the future he'd imagined. But that future was gone now. Eileen had a husband and a son and a life that didn't include Dick.

He thought about her offer to get engaged before he left. About how he'd refused, wanting to give her a way out, not wanting to tie her to him. He'd been right to refuse. She'd needed that freedom. And she'd used it.

Dick sat in the rain until he was soaked through. Then he went back to the barracks, changed into dry clothes, and lay on his bunk staring at the ceiling.

He'd lost Eileen. But his family was still there. His mother, his father, his sisters. They were waiting for him. That had to be enough. It would have to be enough.

Over the next two weeks, Dick carried the loss. It sat in his chest like a physical weight, but he carried it. He'd carried heavier things. He'd carried Price's death. He'd carried the guilt of surviving. He could carry this too.

Other men in the camp had received similar news. Wives who'd remarried. Girlfriends who'd moved on. The war had fractured so many lives, had left so many pieces that couldn't be put back together the way they'd been before. Dick wasn't alone in his loss. That didn't make it easier, but it made it bearable.

In mid-December, Dick's name was called for transport. A troopship was leaving for Australia in three days. He'd be on it, along with two hundred other servicemen heading home.

The ship was called the Stirling Castle, a converted passenger liner that had spent the war ferrying troops and supplies. Now it was bringing men home. Dick stood on deck as they pulled out of Southampton, watching England recede into the grey distance.

He'd come to this country four years ago full of naive hope and adventure. He was leaving it scarred and worn, having learned more about suffering and survival than any man should have to learn.

The voyage took six weeks. Six weeks of grey ocean and grey sky, the ship rolling through swells, heading south through the Atlantic, around Africa, across the Indian Ocean toward Australia.

Dick spent most of his time on deck, watching the water. There were two hundred men on board, but he spoke to few of them. Some were wounded, still healing from injuries sustained in Europe or North Africa. Some were POWs like him, released from camps. Some were just exhausted, war-weary men who'd done their time and wanted nothing more than to go home.

At night, Dick lay in his bunk and listened to the ship creak and groan. He still had nightmares. Price falling. The clearing. The shots. The punishment cells. Werner's face. Beck's shaking hands. Karl and Anna risking everything to help him. All of it mixing together in his dreams until he couldn't tell what was memory and what was invention.

He wrote letters home, though he knew they wouldn't arrive before he did. Writing helped organize his thoughts, helped him process everything that had happened.

Dear Mum,

I'm on a ship heading home. Six weeks at sea, they tell us. Should arrive in Sydney sometime in late January.

I received your letter about Eileen. Thank you for telling me before I arrived. I won't pretend it doesn't hurt, but I understand. She couldn't wait forever. None of you could. You all had to keep living while I was gone.

I think about Price sometimes. The man who died during our escape attempt. I think about how he wanted so desperately to go home, and how that desperation killed him. I think about how I made it and he didn't, and I can't find any justice in that. It was just luck. Timing. Chance.

I think about Tom and the others still in the camps. I hope they're liberated soon. I hope they survive to see the end of the war. But I don't know if they will. So many don't.

I'm different than I was. I know you'll see it when I arrive. The war changed me. I'm thinner, harder, quieter. I don't know if I'll ever be the person I was before. I don't know if that person still exists.

But I'm coming home. That's what matters. I'm coming home and I'll do my best to fit back into life, to be a son and a brother again, to find some kind of normal.

I'll see you soon.

Dick

He never sent that letter. He tore it up and threw the pieces overboard, watching them scatter on the wind. Some things were better left unsaid.

Other men on the ship talked about their plans. Going back to their old jobs. Marrying their sweethearts. Buying houses. Starting businesses. They spoke with enthusiasm, with hope, as if the war hadn't changed everything fundamentally.

Dick couldn't imagine any of those things. Couldn't picture a normal life anymore. Working a regular job, coming home to a wife and children, living as if the war had never happened. How did you do that? How did you pretend that you hadn't seen what you'd seen, hadn't done what you'd done, hadn't lost what you'd lost?

A man named Davies, another Australian heading home to Melbourne, sat with Dick on deck one afternoon. They'd struck up an acquaintance, not quite friendship but companionship of a sort.

"You thinking about what comes next?" Davies asked.

"Not really."

"Me neither. Can't seem to imagine it. The future, I mean. It's all just blank."

"Yeah."

"My girl wrote that she's still waiting. Sarah. But I don't know if that's a good thing or a bad thing." Davies lit a cigarette, offered one to Dick. Dick shook his head. "I'm not the same person she fell in love with. What if she doesn't like who I've become?"

"Then she doesn't," Dick said. "But at least you'll know."

"That supposed to be comforting?"

"No. Just true."

Davies laughed without humour. "Fair enough."

They sat in silence after that, watching the ocean roll past. The ship's engines thrummed beneath them. Gulls followed in their wake, crying out, diving for scraps. The world continued, indifferent to their struggles.

Christmas came and went at sea. The ship's crew put up decorations and served a special meal, turkey and pudding, as much as any man could eat. Dick forced himself to eat, though his stomach rebelled. His body still didn't trust abundance, still expected every meal to be his last.

The men sang carols that night, their voices rough and off-key. Silent Night. O Come All Ye Faithful. Songs that belonged to another world, a world of peace and home and families gathered around warm fires. Dick mouthed the words but didn't sing. His voice had forgotten how.

He thought about last Christmas, in the punishment cell after his second escape attempt. About the cold and the hunger and the absolute certainty that he would die in that camp. He'd been wrong. He'd survived. He'd made it out. But Price hadn't. Tom hadn't. All the others were still there, still waiting, still hoping for liberation.

Dick's survival felt like theft. He'd stolen his freedom from men who deserved it just as much. He'd stolen his chance at life while others remained trapped. The guilt of it sat in his chest beside the loss of Eileen, beside the trauma of the escapes, beside all the other weight he carried.

In early January, the ship crossed into Australian waters. The temperature rose, the grey skies giving way to blue. Men crowded the deck, talking excitedly, pointing at the coastline visible in the distance. Home. After years away, home.

Dick stood at the rail. He should be excited. Should be joyful. But all he managed was a dull awareness: nearly there.

On January 24th, 1945, the Stirling Castle entered Sydney Harbour. Dawn was breaking, mist lying over the water, the city rising behind it. The Harbour Bridge stood silhouetted against the lightening sky. Ferries moved across the water, their whistles echoing. People gathered at the docks, even at this early hour, waving flags, searching the faces on deck for someone they knew.

The men at the rail cheered. Some were crying. Some were laughing. Dick stood among them, watching the scene as if from a distance. This was supposed to be a moment of triumph, of joy, of relief. But all Dick managed was tired.

They docked at Garden Island. The gangway lowered and the men filed off, their boots loud on the wooden dock. Australian soil. Home soil. Dick stepped onto it and waited for something to shift inside him. Anything. But the numbness remained.

Processing took hours. They lined up for medical inspections, paperwork, and pay. Clerks in uniform checked names against lists, stamped forms, and handed out envelopes. Dick's pay envelope was thick with money he'd earned but never spent, accumulating over years of captivity. He shoved it in his pocket without counting it.

A doctor checked his chest, his reflexes, and his shoulder. Made notes in a file. Asked routine questions. Dick answered mechanically. Yes, he was eating. Yes, he was sleeping. No, he didn't need immediate medical attention.

The doctor looked at him with knowing eyes. "You've been through something terrible. It's going to take time to adjust. Be patient with yourself."

Dick nodded. More waiting. Everything was waiting.

They were given the option to stay at barracks near the docks or continue their journey home immediately. Dick chose to stay one night. He wasn't ready yet.

Wasn't ready to face his family, to step into the role of son and brother, to pretend he was whole when he felt broken.

That night, lying on a bunk in barracks that smelled of eucalyptus and sea air, Dick listened to the other men talk. They were planning their homecomings, their reunions, their futures. Their voices were full of hope and excitement and relief.

Dick closed his eyes and tried to access those things. Tried to find the joy that should be there, the gratitude, the peace. But all he found was emptiness and a fear that he'd forgotten how to be human, how to be anything other than a prisoner.

He pulled out the photograph one more time. His family. They were so close now. Just a train ride away. In less than a day, he'd see them. His mother's face. His father's handshake. His sisters" voices. Everything he'd fought for, everything he'd survived for, everything he'd carried this photograph through hell to get back to.

Dick put the photograph away and tried to sleep. Tomorrow. Tomorrow he'd go home. Tomorrow he'd face whatever came next.

Tomorrow.

The next morning, Dick boarded the train to Newcastle. It pulled out of Central Station slowly, moving past rows of red roofs and backyards where washing hung in the morning sun. Beyond the city, the landscape opened up. Gum trees, so distinctly Australian, their grey-green leaves shimmering in the heat. The red earth. The wide sky. The sweep of the coastline visible in the distance.

Dick sat by the window and watched it all pass. This

was his country. This was home. He'd dreamed of this landscape in the camps, had held onto these images when everything else faded. And now he was here, really here, and it felt surreal.

Other passengers glanced at him curiously. His uniform marked him as returned military. An older woman smiled at him kindly. "Welcome home, love," she said.

"Thank you," Dick replied. His voice sounded strange to his own ears.

The train rattled through small towns whose names Dick knew by heart. Gosford. Wyong. Morisset. Each one brought him closer to Newcastle, closer to the moment he'd been dreaming of and dreading in equal measure.

He dozed in short bursts, his body exhausted but his mind refusing real rest. Every time he closed his eyes, he saw the camps. The wire. The guards. Price falling. The punishment cells. His mind couldn't let go, couldn't stop replaying the trauma even now when he was safe.

By afternoon, the train reached Newcastle. Dick stepped down onto the platform, his small bag in hand. The station was busy with afternoon traffic. People hurrying past, carrying packages, meeting loved ones, living their normal lives.

Dick stood still for a moment, overwhelmed by the noise and the movement and the sheer normalcy of it all. This was what he'd been fighting to get back to. This ordinary afternoon in an ordinary city. But it felt alien now, like a world he no longer belonged to.

He walked out of the station, following streets he knew by memory. The air smelled of salt and coal dust from the port. The sky was blue and huge and empty of bombers. The sun was hot on his shoulders. Australia in summer. Everything was the opposite of the Polish winter he'd escaped from.

Dick walked toward Alfred Street, toward the house where his family had lived, where the photograph had been taken. He needed to see it, even though he knew they'd moved. He needed to touch that piece of his past before moving forward.

The house looked smaller than he remembered. The paint was faded, the verandah weathered by years of sea air. A different family lived there now. He could hear children playing in the backyard, their voices high and carefree.

Dick stood across the street, looking at the house, remembering. The front verandah where he'd sat with his mates talking about the war. The backyard where the photograph was taken. The front door he'd walked through a thousand times. It was all still there, but it wasn't his anymore. That life was gone.

He walked to the front door and knocked. A man he didn't recognize answered, looking at him with mild curiosity.

"Can I help you?"

"I'm looking for the Roberts family," Dick said. "They used to live here."

"They moved on some time back," the man replied. "Year or so ago, I think. Left a note in case anyone came asking." He disappeared for a moment and returned with a small envelope. "Here you go."

Dick took the envelope with shaking hands. Inside, in his mother's neat handwriting, were a few lines:

We are now at 132 Brunker Road, Broadmeadow. Come find us.

Dick thanked the man and walked back toward Hunter Street. His legs felt weak. This was it. The moment he'd been imagining for three years. The homecoming. The reunion. Everything he'd suffered for was about to become real.

He caught a tram heading west, paid his fare with strange-feeling coins, and sat by the window watching the familiar streets pass. The tram bell clanged. The wheels rattled on the tracks. The sound and feel of it made his chest tighten with emotion he couldn't name.

The journey took twenty minutes. Dick stepped off at the corner of Brunker Road and stood still for a moment, looking at the wide street lined with pepper trees. The afternoon sun filtered through the leaves, dappling the footpath. A warm breeze carried the scent of eucalyptus.

Number 132 was easy to find. A weatherboard house with a front verandah and a large yard behind it. Bigger than the Alfred Street house, more room for everyone. Dick could see washing on the line, the shapes of his mother's dresses and his sisters" blouses stirring in the breeze.

He walked up to the gate and stopped. His hands were shaking. His pulse pounded so hard in his temples he could hear it. This was it. This was the moment. Three years of suffering, three escape attempts, Price's death, everything, all of it had led to this moment.

Dick opened the gate and walked up the path. Before he reached the front steps, the door opened.

His mother came out first.

She was older. That was Dick's first thought. She'd aged in the three years he'd been gone. Her hair was greyer. Her face was more lined. But her eyes were the same, warm and steady, and they filled with tears the moment she saw him.

"Dick," she said. Just his name. Nothing more.

Then she was moving toward him, down the steps, across the yard, and Dick was moving too, his bag dropping to the ground, his arms opening. They met halfway and she wrapped her arms around him and Dick realized he was crying, tears running down his face, his body shaking with sobs he couldn't control.

"You're home," his mother said, her voice thick with emotion. "You're home, you're home, you're home."

Then his sisters were there. Jean first, married now, fuller in the face, her hand seeking Dick's. Then Lorna, thinner than he remembered, her sharp features softened by tears. Then Dorrie and Shirley and Meryl, all of them older, all of them changed, all of them surrounding him, touching him, crying and laughing at the same time.

His father stood on the verandah, his hands in his pockets, his face working to control emotion. When Dick looked at him, when their eyes met, something passed between them. Understanding. Recognition. Man to man, father to son. His father came down the steps slowly and extended his hand.

Dick took it. They shook, firm and steady. Then his father pulled him into an embrace, brief but fierce.

"Welcome home, son," his father said, his voice rough.

Dick couldn't speak. His throat was too tight, his chest too full. He just nodded and let himself be led inside, surrounded by his family, by voices talking over each other, by hands touching his arms and his back as if to prove he was real.

Inside, the house smelled of home. Tea brewing. Wood polish. Warm bread. The particular scent of his family's life continuing while he'd been gone. Everything was different from Alfred Street, but everything was also the same. The sounds. The warmth. The love.

They sat him at the kitchen table, and Dick looked around at the faces surrounding him. His mother making tea, her hands steady despite the tears on her cheeks. His father pulling out a chair and sitting beside him. His sisters all talking at once, asking questions, telling him news, filling the space with sound.

This was what he'd dreamed of. This exact moment. The kitchen table, his family around him, the sense of being home. He'd carried the photograph of this through camps and escapes and punishment cells, and now he was living it.

It was overwhelming. Too much sensation, too much emotion, too much everything after years of deprivation. Dick felt like he was drowning in it, unable to process it all, unable to be fully present.

His mother set tea in front of him and he wrapped his hands around the cup, feeling its warmth. Real tea, strong and sweet. He drank it slowly and let the conversation wash over him.

Jean was telling him about her wedding. Lorna was complaining about her job at the shop. Dorrie was talking about her teacher training. Shirley and Meryl were arguing about something trivial. His father was asking careful questions about the journey home, avoiding the war itself.

Dick answered when spoken to directly, but mostly he just listened. Just absorbed the reality of being here, of being with them, of being home.

As evening fell, his mother served dinner. Roast lamb and vegetables and fresh bread. More food than Dick had seen in three years. His plate was piled high and he stared at it, unable to believe the abundance.

"Eat, darling," his mother said softly. "You need to put some weight on."

Dick ate slowly, forcing himself not to rush, not to make himself sick. His body still didn't trust food, still expected every meal to be his last. But he ate, and it was delicious, and his mother smiled to see him eating her cooking again.

After dinner, as his sisters cleared the table and his mother served tea, Dick's father pulled him aside to the front verandah. They stood in the warm evening air, listening to the sounds of the neighbourhood settling for the night.

"You've been through hell," his father said quietly. "I can see it on your face."

Dick nodded.

"You don't have to talk about it if you don't want to. But

if you do, I'll listen. Your mother will listen. We're here."

"Thank you," Dick said. His voice was rough.

"Take your time settling back in. There's no rush. You need time to heal, to rest, to remember what normal life feels like. We understand that."

Dick looked at his father, at this man who'd always been reserved and stern but who was showing such gentleness now. "I'm not the same person who left."

"None of us are. The war changed all of us, even those of us here at home. But you're still our son. Still our brother. That hasn't changed."

Dick nodded, unable to speak around the lump in his throat.

That night, they showed him to his room. It was small but clean, with a window overlooking the backyard. His mother had made up the bed with fresh sheets. A dresser stood against one wall. A chair by the window.

"It's yours for as long as you need it," his mother said. "There's no rush to find your own place, to get a job, to do anything. You just focus on getting well."

After everyone had gone to bed, Dick lay in his own room, in his own bed, and stared at the ceiling. The house was quiet around him. He could hear the wind moving through the trees outside, could smell the salt air from the coast, could feel the soft mattress beneath him.

He was home. He'd actually made it. After everything, after all the suffering and the fear and the loss, he was here. In Newcastle. In his family's house. Safe.

Dick pulled out the photograph one last time. He held it

up in the moonlight coming through the window and looked at the faces. This was where the photograph had been taken. In this backyard, or the one at Alfred Street. This was where his family had stood, frozen in time, waiting for him.

He'd brought the photograph home. He'd brought himself home. The circle was complete.

Dick set the photograph on the dresser and lay back down. His body was exhausted. His mind was overwhelmed. But underneath all of that, buried deep but present, was something that might have been peace.

He closed his eyes and listened to the wind in the trees. It sounded almost like the sea. Like the surf at the beach he'd dreamed about for three years. Like home.

Tomorrow would bring its own challenges. Tomorrow he'd have to figure out how to live in this world again, how to be a person instead of a prisoner, how to carry forward the weight of what he'd experienced.

But tonight, he was home. And that was enough.

CHAPTER 20— AFTER

The first night, Dick slept for twelve hours. When he woke, the sun was high and the house was quiet. For a moment he didn't know where he was. The softness of the mattress, the clean smell of the sheets, the warmth of the room, all of it felt wrong. His body tensed, waiting for the whistle, for the guards, for the cold reality of the camp.

Then he remembered. Home. He was home.

Dick sat up slowly, his shoulder aching with the movement. Through the window he could see the backyard, green and peaceful. A magpie was singing its morning song. The wind rustled through the pepper trees. Normal sounds. Safe sounds.

He dressed in civilian clothes his mother had laid out for him. They hung loose on his frame, several sizes too big. His mother must have guessed at his size, not knowing how much weight he'd lost. Dick rolled up the sleeves and cinched the belt tight and looked at himself in the small mirror on the dresser.

The face looking back was gaunt and hollow-eyed. His cheekbones were sharp. His neck was thin. He looked ten years older than his twenty-six years. This was what the war had made him. This was what survival looked like.

Dick went downstairs and found his mother in the kitchen. She smiled when she saw him, but he caught the flicker of concern in her eyes.

"Good morning, darling. I let you sleep. You needed it. Are you hungry?"

"Yes."

She made him breakfast. Eggs and bacon and toast with butter and jam. More food than Dick would have seen in a week at the camp. He ate slowly, methodically, his body still not trusting the abundance.

His mother sat across from him with her tea, watching him eat with quiet satisfaction. "You'll fill out again," she said. "It'll just take time."

"I know."

"Your sisters have gone to work. Jean and Lorna both. Dorrie's at her training. Shirley and Meryl are at school. It'll be quieter today. You can rest more if you need to."

Dick nodded. He didn't know what to do with himself. For three years, every moment of his day had been scheduled, controlled, dictated by guards and whistles and routines. Now he had freedom and didn't know what to do with it.

"What would you like to do today?" his mother asked gently.

"I don't know."

"You could walk down to the beach if you like. You always loved the beach."

The beach. Dick had dreamed about it for three years. The sand, the surf, the smell of salt air. "Yeah," he said.

"Maybe I will."

After breakfast, Dick walked. Not to the beach, not yet. Just through the streets of Broadmeadow, relearning the geography of home. Everything was familiar but strange. The houses, the shops, the people going about their daily lives. They didn't know what he'd been through. They couldn't see the camps in his eyes, couldn't hear the screams in his head.

He was a ghost walking through a world that had continued without him.

A woman passed him on the footpath and smiled. "Lovely day, isn't it?"

"Yes," Dick replied automatically.

Lovely day. The words felt meaningless. What was lovely about it? The sun? The blue sky? The fact that no one was shooting at him? Dick didn't know how to measure lovely anymore. His scale had been recalibrated by suffering.

He walked until he was tired, then walked some more. His body was weak from months of starvation and hard labour. His shoulder ached constantly. But the movement felt good, felt purposeful. He was choosing where to go, when to stop, whether to turn left or right. Simple choices that had been taken from him for so long.

When he returned home, his father was there, sitting on the front verandah reading the newspaper. He looked up as Dick approached.

"Have a good walk?"

"Yes."

"Getting reacquainted with the place?"

"Yeah."

His father folded the newspaper and set it aside. "There's no rush to figure everything out. No rush to find work or make plans. You take whatever time you need."

Dick sat down in the other chair. "I don't know what to do with myself."

"That's understandable. You've been living under orders for years. Freedom takes adjustment." His father paused. "Your mother and I, we're just glad you're home. However long it takes you to settle, we're here."

"Thank you."

They sat in silence for a while, father and son, both looking out at the street. His father had always been a man of few words, practical and steady. He'd worked at the Herald for decades, setting type, coming home with ink on his hands. He didn't try to understand what Dick had been through, didn't pretend to have answers. He just offered his quiet presence, and that was enough.

That night, Dick had his first nightmare at home. He was back in the punishment cell, cold and dark and alone. He was running through the snow while shots cracked around him. He was watching Price fall, blood spreading dark across the white ground.

He woke up gasping, his body drenched in sweat. For a moment he didn't know where he was. Then his mother was there, sitting on the edge of his bed, her hand on his shoulder.

"You're safe," she said softly. "You're home. You're safe."

Dick's breathing slowly returned to normal. His mother didn't ask what he'd been dreaming about. She just sat with him until he was calm, then brought him water and stayed until he fell back asleep.

It became a pattern. Nightmares most nights. His mother or one of his sisters coming to sit with him, to remind him where he was, to anchor him to the present. They never complained. Never made him feel like a burden. They just accepted it as part of having him home.

The days passed slowly. Dick walked a lot, exploring Newcastle, relearning the city he'd grown up in. He went to the beach finally, stood on the sand and looked out at the ocean. The same water that touched Europe, that had surrounded the ship bringing him home, that had always been here waiting for him.

He took off his shoes and walked into the surf, letting the waves wash over his feet. The cold shock of it was grounding. Real. This was real. He was really here.

A few weeks after arriving home, Dick saw Eileen.

He was walking through town, running an errand for his mother, when he saw her coming out of the pharmacy. She was pushing a pram, a small baby bundled inside. A man walked beside her, tall and well-dressed, his hand on her elbow. The husband. The doctor.

Eileen looked happy. That was what struck Dick most. She was laughing at something her husband had said, her face bright and open. She looked like someone who'd moved on, who'd built a life, who wasn't haunted by the past.

Their eyes met for a moment across the street. Recognition flickered in her face. Surprise. Then something complicated, guilt maybe, or sadness, or just the weight of shared history.

Dick nodded once. A small acknowledgment. I see you. I understand. It's all right.

Eileen nodded back. Then her husband said something and she turned away, pushing the pram down the street, her attention back on her present instead of her past.

Dick stood there for a moment, watching her go. Loss, yes. Grief for what might have been. But also acceptance. Eileen had made the right choice. She had a husband who loved her, a baby, a life. She was happy. That was more than Dick could have given her, damaged and broken as he was.

He turned and walked the other way, back toward home. That chapter was closed. Eileen belonged to his past, to the person he'd been before the war. She couldn't be part of his future because that future was still being written, still uncertain, still taking shape.

By March, Dick needed to find work. Not for the money, his back pay and the small pension for his injured shoulder would cover his basic needs. But for purpose. For structure. For something to do with his days besides walk and think and remember.

He tried the docks first. The hiring manager looked at his thin frame and injured shoulder and shook his head. "You're not strong enough, mate. Not yet. Maybe in a few months when you've put some weight on."

Dick tried the steelworks next. Same response. He was

too thin, too obviously unwell, too much of a liability.

Finally, in April, he found work as a wool classer at the sheds in Carrington. The foreman, an older man named Bill, looked Dick over and saw past the gaunt frame to something else.

"You're one of the boys who made it back," Bill said. It wasn't a question.

"Yes."

"Prisoner?"

"Yes."

Bill nodded slowly. "I can give you a go. Work's not too heavy. Sorting and grading wool. Pays decent. You show up on time and do the work, we'll get along fine."

"Thank you."

"Don't thank me yet. See how you go first."

The work was repetitive but honest. Dick sorted through fleeces, grading them by quality, separating them into bins. His hands learned the feel of different grades of wool. His body adapted to the rhythm of the work. Eight hours a day, Monday through Friday. Structure. Purpose. Normalcy.

The other workers were a mix of ages. Some were older men like Bill. Some were younger men, either too young for service or classified as essential workers. They didn't ask Dick about the war. They could see he'd been through something, and they respected the silence around it.

At lunch, Dick would sit on the dock and eat the sandwiches his mother had packed. He'd watch the ships moving in the harbour, loading and unloading cargo, going about their business. The world turning, indifferent to his struggles.

Bill would sometimes sit with him, not saying much, just being present. Once, he said, "It gets easier. Not better, exactly. But easier to carry."

"I hope so."

"It will. It just takes time." Bill lit his pipe. "You keep showing up, keep doing the work, and one day you'll realize you've built something. A life. That's all any of us can do."

By winter, Dick had gained back some weight. His clothes fit better. His face had filled out slightly. He still looked older than his years, still had the haunted look in his eyes, but he was starting to resemble a living person instead of a ghost.

News came through the Red Cross that Stalag VIII-A had been liberated in March. The prisoners were free. Dick read the notice and felt a complicated mix of emotions. Relief that Tom and Walsh and the others had survived. Guilt that he'd gotten out early while they'd endured months more of captivity. Grief for all the time lost, for all the suffering that could never be undone.

He wrote to Tom, sending the letter through military channels, hoping it would find him. He wanted Tom to know he'd made it home, that the escape had been worth it, that someone had gotten out.

Dear Tom,

I heard the camp was liberated. I hope you're all right. I hope you're home or on your way home soon.

I made it to Australia. I'm in Newcastle, living with my family, working at the wool sheds. I'm managing. Getting by. Learning how to be a person again instead of a prisoner.

Thank you for giving me Miller's map. Thank you for trying to save me from myself. Thank you for being my friend when I needed one most.

I think about Price sometimes. About what he wanted and what it cost him. I carry that with me. I'll always carry it.

Take care of yourself. Come visit if you ever make it to Newcastle.

Dick

He never received a response. He didn't know if the letter reached Tom, or if Tom was even alive to receive it. That uncertainty was hard, but Dick had learned to live with uncertainty. The war had taught him that certainty was an illusion anyway.

The seasons turned. Winter gave way to spring. Spring gave way to summer. Dick kept working, kept living, kept putting one foot in front of the other. He rarely spoke about the war. When people asked, he'd give brief answers and change the subject. Only his family knew the shape of what he'd been through, and even they didn't know the details.

Some things were too heavy to share. Some experiences couldn't be put into words. Dick carried them alone, locked inside, where they couldn't hurt anyone else.

In late 1947, Dick met Joy.

She worked at the library in Newcastle, a cheerful

woman with warm brown eyes and an easy laugh. Dick had started going to the library on weekends, looking for something to occupy his mind, and Joy would help him find books.

"You like history," she observed one Saturday afternoon, looking at his selections.

"I suppose I do."

"Any particular period?"

"Anything before 1939," Dick said, then caught himself. "Sorry. That sounded bitter."

"No, I understand." Her eyes were kind. "You were in the war."

"Yes."

"It must have been difficult."

"It was."

She didn't press for details. She just smiled and stamped his books. "These are due back in two weeks. Enjoy them."

Over the following months, Dick found himself timing his library visits to when Joy was working. They'd talk briefly while she checked out his books, conversations that slowly grew longer, more personal. She had a lightness about her that Dick found soothing. She could make him smile, something he'd almost forgotten how to do.

It took him four months to ask her to dinner. She said yes without hesitation, her face lighting up.

"I was wondering if you'd ever ask," she said. "I've been hoping you would."

They had dinner at a small restaurant near the harbour. Dick was nervous, his hands shaking slightly as he held the menu. It had been so long since he'd done anything like this. Since he'd tried to be a person instead of a survivor.

But Joy made it easy. She talked about books and her work at the library and her family. She asked him questions but never pushed when he didn't want to answer. She laughed at his awkward jokes and made him feel like maybe, just maybe, he could have this. Could have someone to share his life with.

"I should tell you," Dick said as they walked along the harbour after dinner, "I'm not... I'm not whole anymore. The war, it changed me. I have nightmares. I'm quiet. I don't talk about what happened. I might never talk about it."

Joy stopped walking and looked at him. "Do you think that matters to me?"

"It should."

"Well, it doesn't." She took his hand, the first time she'd touched him. "Everyone has something, Dick. Everyone carries weight. You carry yours from the war. That's all right. I'm not asking you to be perfect. I'm just asking if you'd like to have dinner again sometime."

Dick looked at her, at this woman who saw his damage and didn't flinch, who offered acceptance instead of demands.

"Yes," he said. "I'd like that very much."

They courted for a year, taking things slowly. Joy met his family and charmed them all, especially his mother who'd been worried Dick would never find someone. Dick met Joy's parents, who welcomed him warmly despite his obvious struggles.

In 1948, Dick proposed. They were at the beach, his favourite place, watching the sunset paint the water gold and pink.

"I don't have much to offer," Dick said. "Just a job at the wool sheds and a small flat I'm renting. But I love you. And I'll spend the rest of my life trying to be worthy of you."

Joy kissed him. "You're already worthy. And yes, I'll marry you."

They married in 1949 at St Peter's. It was a simple ceremony, just family and close friends. Dick wore a suit that finally fit properly, his body having recovered its strength. Joy wore a white dress and carried flowers from his mother's garden.

As they said their vows, Dick thought briefly about Eileen. About what might have been. But that belonged to a different life, a different version of himself. This was his reality now. Joy, warm and steady, willing to build a life with a damaged man.

They rented a flat in Carrington at first, near the wool sheds where Dick worked. Joy continued at the library part-time. Together they started building something that resembled normal life.

Their first son was born in 1947. Another son in 1949. Their first daughter came in 1953.Their youngest daughter in 1955.

Dick was a good father. Quiet, but present. He taught his

sons to fish and his daughters to swim. He helped with homework and attended school concerts. He provided and protected and did all the things fathers were supposed to do.

But there was always a distance. A part of him that couldn't fully engage, couldn't fully be present, because part of him was still in those camps. Still running through the snow. Still watching Price fall. Still locked in that punishment cell.

Joy understood. She never complained about the distance. She just worked around it, giving him space when he needed it, drawing him back when he drifted too far. She'd wake with him when the nightmares came, holding him until the shaking stopped, never asking what he'd seen.

By the mid-1950s, Dick and Joy had saved enough to buy a house. They found one in New Lambton, a solid brick home with a backyard big enough for the children to play. It needed work, but Dick spent his weekends fixing it up, repairing and painting and making it theirs.

The house became their anchor. A place that was fully theirs, earned through years of saving and hard work. Joy planted a garden. Dick built a workshop in the garage where he could tinker with things. The children grew up there, their heights marked on the kitchen doorframe, their laughter filling the rooms.

The children knew their father had been in the war, but they didn't know the details. Dick never spoke about it. When they asked, he'd say something vague and change the subject. The war was a closed door in their household. Something acknowledged but never opened.

Only once did Dick talk about it in any detail. His brother-in-law Alistair, Jean's husband, was visiting in the late 1950s. They were sitting on the verandah after dinner while the women cleaned up and the children played in the yard.

Alistair lit a cigarette and offered one to Dick. Dick shook his head.

"You were a prisoner," Alistair said. "Poland, wasn't it?"

"Yes."

"That must have been rough."

"It was."

Alistair was quiet for a moment. Then he said, "I heard you tried to escape."

"Three times." Dick's voice was flat.

Alistair's eyebrows rose. "Three times? That takes guts."

"Or stupidity. It's hard to tell the difference sometimes." Dick paused. "A mate died during one attempt. Price. We tried together. He got shot. I ran."

"That's rough."

"The third time I made it. I walked to Switzerland. I took help from people who could have been killed for helping me." Dick's voice was quiet. "I think about that sometimes. About the cost. About whether any of it was worth it."

"You made it home. That's worth something."

"Others didn't."

Alistair didn't have an answer for that. They sat in

silence, smoking, watching the evening darken around them.

Dick kept working at the wool sheds through the years. The work was steady, reliable, something he could count on. Bill retired in the early sixties, and a younger foreman took over, but the work remained the same. Sorting fleeces, grading wool, the same rhythm day after day, year after year.

His children grew. His sons went into trades, working with their hands like their father. His daughters married and started families of their own. The house in New Lambton filled with grandchildren on weekends, their noise and energy both exhausting and life-giving.

Dick aged alongside them, his hair going grey, his body slowing. The shoulder that had been shot never fully healed. It ached in cold weather, a permanent reminder of that desperate run across the field.

He retired from the wool sheds at sixty-five, his body finally demanding rest. Retirement was strange. Dick didn't know what to do with unlimited time. He'd spent so much of his life working that freedom was still uncomfortable.

He took to walking in the mornings, down to the beach when he could, or just through the neighbourhood. Joy would sometimes come with him, the two of them walking in comfortable silence, their hands occasionally touching.

The photograph, the one he'd carried through everything, sat in a frame on the mantelpiece. Faded and worn, the blood stains barely visible anymore, but treasured. His children would sometimes ask about it, about why it looked so old and damaged.

"I carried it through the war," Dick would say. "It reminded me of what I was fighting to get back to."

That was all he'd say. The details, the camps, the escapes, the punishment cells, Price's death, all of that stayed locked inside.

In 1984, Dick had a stroke. He was seventy years old, working in the garden at the house in New Lambton, when he felt a sudden pressure in his head. Joy found him collapsed among the tomato plants and called for help.

He survived, but he was never quite the same. His left side was weakened. His speech was slurred. The doctors said he was lucky it hadn't been worse, but Dick didn't feel lucky. His body had finally given out, surrendered after decades of carrying trauma.

Joy cared for him with the same patience she'd shown throughout their marriage. She helped him with physical therapy, helped him relearn basic tasks, and never complained about the burden. The children visited regularly, helping where they could.

But Dick was fading. They could all see it. The stroke had taken something essential from him, some core strength that had kept him going all these years. He slept more. He talked less. Seemed to be withdrawing from the world.

In early 1985, Dick had another stroke. Massive this time. He died in the hospital with Joy holding his hand, his children gathered around the bed.

"He was a good man," the priest said at the funeral. "A

good husband, a good father, a good friend. He survived something terrible in the war and never let it make him cruel. That's what we should remember."

They buried him in Sandgate Cemetery, not far from the beach he'd loved. The headstone was simple:

James Owen Roberts

1914-1985

Beloved Husband and Father

He Came Home

After Dick's death, Joy couldn't stay in the house in New Lambton. It was too full of memories, too empty without him. Her youngest daughter invited her to move in with her family, and Joy accepted gratefully.

She lived with them for the next ten years, helping with the grandchildren, keeping busy, but always carrying the absence of Dick like a physical weight. She kept the photograph with her, on her bedside table, looking at it every night before she slept.

She died peacefully in her sleep in 1995, at the age of seventy-six. Her daughter found her in the morning, looking peaceful, as if she'd simply decided it was time to join Dick.

They buried her beside him. Two people who'd found each other after the war, who'd built something good despite the damage, who'd loved each other for nearly forty years.

At the wake, the children told stories about their parents. About their father's quiet strength. About their mother's endless patience. About the life they'd built together in the house in New Lambton.

"Dad never talked about the war," his eldest son said. "But you could see it in him. In the nightmares. In the way he'd startle at loud noises. In the distance in his eyes sometimes."

"Mum understood him," his daughter added. "She knew how to reach him when he drifted away. She never gave up on him, even when it must have been hard."

"He survived something terrible," his younger son said. "And he never let it destroy him. That's what I'll remember most. He had every reason to be broken, but he chose to keep living. To build a family. To love us all."

"They were good parents," his youngest daughter said simply. "We were lucky to have them."

The family kept the photograph, passing it down through generations. Dick and Joy's grandchildren would ask about it, about the blood stains and the worn edges, about why it was so treasured.

And the children would tell them. About Dick carrying it through prison camps. About him using it to remember who he was, what he was fighting to get back to. About how he made it home despite everything, and how he built a good life despite the damage.

The photograph became a family treasure. A reminder of survival, of hope, of the power of love and home to sustain a person through the darkest times.

Dick Roberts had survived the war. He'd escaped three times. He'd watched friends die. He'd suffered in ways his children would never fully understand.

But he'd come home. He'd found Joy. He'd built a family.

He'd lived a good life in the house in New Lambton, surrounded by the people he loved.

That was his victory. Not the escapes, not the survival, but the life he built after.

That was his triumph.

BOOKS BY THIS AUTHOR

The Debt Collectors Of Auschwitz

Six survivors. One impossible escape. A promise of justice that will follow them across continents.

When the SS begin executing prisoners to hide the evidence of their crimes, six inmates at Birkenau seize their only chance to break free. Led by the fiercely determined Miriam and the haunted neurosurgeon Dr. Levi Blum, the group fight through ice, darkness and gunfire to escape the camp in its final days.

But freedom comes with a vow: they will not walk away from the dead.

From the ruins of post-war Europe to the streets of New York, the survivors hunt the men who destroyed their families. They forge documents, rob banks, and build new identities as they pursue justice on their own terms. Yet the past refuses to stay buried.

When Levi discovers a former Auschwitz doctor living under a false name, he is forced to choose between vengeance and the oath that once guided him. His decision will test the fragile bond that holds the group together.

Gripping, emotional and inspired by real accounts, The Debt Collectors of Auschwitz is a story of survival, courage and the relentless search for truth. For readers of The Tattooist of Auschwitz, Cilka's Journey and The Choice.

Revenge First

When everything you love is torn away, how far will you go for justice?

Sixteen-year-old Daniel witnesses the unthinkable: the brutal execution of his entire family. Standing over their graves, he makes a sacred vow. No matter how long it takes, no matter what it costs, he will have his revenge.

His journey to Palestine becomes an odyssey of survival and transformation. Through years of hardship, Daniel evolves from a broken boy into a hardened warrior, learning to fight, to endure, and to live with the fire of vengeance burning in his heart.

But when he finally discovers the whereabouts of his family's killer, fate presents him with an impossible choice. He has found something he never expected to find again: love. Now he must decide. Will he honor the oath that has sustained him through the darkest years of his life, or will he choose the possibility of a future filled with hope?

A gripping tale of loss, resilience, and the price of revenge, "Revenge First" explores the weight of promises made in grief and the transformative power of unexpected love.

www.ingramcontent.com/pod-product-compliance
Lightning Source LLC
Chambersburg PA
CBHW032148080426
42735CB00008B/626
* 9 7 8 1 7 6 4 1 9 6 9 4 9 *